# Engineering Surveying

Volume 1, Third Edition

# Engineering Surveying

## Theory and Examination Problems for Students

Volume 1, Third Edition

## W. Schofield

MPhil, ARICS, FInst CES, Assoc IMinE
Principal Lecturer, Kingston Polytechnic

**Butterworths**
London Boston Durban Singapore Sydney Toronto Wellington

First published 1972
Second edition 1978
Third edition 1984

© W. Schofield, 1984

**British Library Cataloguing in Publication Data**

Schofield, W.
  Engineering surveying.—3rd ed.
  Vol. 1
  1. Surveying
  I. Title
  526.9    TA545

  ISBN 0–408–01227–7

**Library of Congress Cataloguing in Publication Data**

Schofield, W. (Wilfred)
  Engineering surveying.
  Bibliography: p. 000
  Includes index.
  Contents: v. 1. Theory and examination problems
for students.
  1. Surveying.    I. Title.
TA45.S263  1984      526.9′024′62      83–7787

  ISBN 0–408–01227–7 (v. 1)

Filmset by Mid-County Press, London SW15
Printed and bound in England by Redwood Burn Ltd, Trowbridge, Wilts.

# Preface

This book specializes in surveying for engineers, treating the subject as field metrology.

The emphasis throughout is to develop in the reader a clear understanding of the basic concepts of each topic in order that they may be correctly utilized and applied to the many different 'on-site' situations as they arise. To reinforce these concepts, each chapter contains a great many 'worked examples', carefully selected from a wide variety of representative examination sources, followed by student exercises complete with answers.

Throughout the book, various topics are dealt with in relation to modern surveying techniques and instrumentation, and with due regard to the impact of information technology on data capture and processing.

*Chapter 1:* a new chapter dealing with the overall basic concepts of surveying procedure and showing the application of the fundamental measure of angle and length in the surveying process. It utilizes a simple model to illustrate the importance of reconnaissance; of working from the whole to the part; of the careful consideration of all error sources and independent checks.

*Chapter 3* (Earthworks): has been expanded and includes an extremely useful program, written in basic, for the computation of areas, volumes, mass-haul ordinates and slope stake cutting points, in route construction.

*Chapter 4:* completely reorganized and containing much new material on centring errors and errors affecting steep sights, which are of importance to the engineering surveyor.

*Chapter 5* deals with optical distance measurement and concentrates on those areas which it is felt will not be rendered obsolete by the advent of EDM. Also, the reader is introduced to the standard treatment of small errors by partial differentiation.

*Chapter 6* deals with the design and setting out of circular, transition and vertical curves plus the principle and use of the osculating circle and the very latest design procedures which, at the time of writing, are not yet available in other texts.

*Chapter 7:* a unique chapter dealing exhaustively with the specialist procedures required by the tunnelling and mining engineer. It also includes an introduction to hydrographic surveying, sufficient for most in-shore surveys required of engineers.

*Chapter 8:* another new chapter which extensively covers the many varied aspects of setting out (dimensional control) on site. It includes much detail on the application of lasers, the law and hazards appertaining to their use and the responsibilities and organization required on site to ensure prompt, accurate and economic setting-out.

Whilst the book is a work of reference for practitioners it is also aimed at easy student assimilation of theory and its eventual application in the field; thus the importance of examination success in the subject has not been overlooked. A careful study of the worked examples and diligent completion of all the set exercises will equip the student for this task.

The book should prove useful not only to technician and undergraduate students of surveying, civil mining and municipal engineering, but also to those studying for the various professional examinations which cover this subject.

## Acknowledgements

I am indebted, in the first instance, to my colleague Mr B. J. Merrony, BA, MSc, DIC, for his kind permission to include his excellent earthworks program, which I feel sure will be utilized by, and of great benefit to, all who read this book.

I am also indebted to the Controller of HM Stationery Office and the County Surveyors' Society for permission to reproduce the various curve-design and setting-out tables. Also to the Senate of London University, Kingston-upon-Thames Polytechnic and the Institution of Civil Engineers, for their permission to reproduce questions set in their examinations.

W. Schofield

*To my wife Jean and daughter Zoë*

# Contents

# Basic concepts of surveying

The aim of this Chapter is to introduce the reader to the basic concepts of surveying. It is therefore the most important Chapter and worthy of careful study and consideration.

## 1.1 DEFINITION

Surveying may be defined as 'the science of determining the position, in three dimensions, of natural and man-made features on or beneath the surface of the Earth'. These features may then be represented in analogue form as a contoured map, plan or chart, or in digital form as a three-dimensional mathematical model stored in the computer. This latter format is referred to as a *digital ground model* (DGM).

In engineering surveying, either or both of the above formats may be utilized in the planning, design and construction of works, both on the surface and underground. At a later stage, surveying techniques are used in the dimensional control or setting out of the designed constructional elements and also in the monitoring of deformation movements.

In the first instance, surveying requires management and decision-making in deciding the appropriate methods and instrumentation required for satisfactory completion of the task to the specified accuracy and within the time limits available. Only after very careful and detailed *reconnaissance* of the area to be surveyed can this initial process be properly executed.

When this is complete, the field work—involving the capture and storage of field data—is carried out with the aid of instruments and techniques appropriate to the task in hand.

The next step in the operation is that of data processing. Most, if not all, of the computation will be carried out on computers which may range in size from pocket calculators to large main-frame computers. The methods adopted will depend upon the size and precision of the survey, as well as the manner of its recording whether this be in a field book or a data logger.

Data representation in analogue or digital form may now be carried out by conventional cartographic plotting or through a totally-automated system using a computer-driven flat-bed plotter.

In engineering, the plan or DGM is used for the planning and design of a construction project, which might be a railroad or a highway, dam, bridge, or even a new town complex. No matter what the work is, or how complicated, it must be set out on the ground in its correct place and to its correct dimensions, within the tolerances specified. To this end, surveying procedures and instrumentation of varying precision and complexity are used, the choice depending upon the nature of the project in hand.

Since surveying is indispensable to the engineer in the planning, design and construction of a project, all engineers should have a thorough understanding of the limits of accuracy possible in the construction and manufacturing processes. This knowledge, combined with an equal understanding of the limits and capabilities of surveying instrumentation and techniques, will enable the engineer to complete his project successfully in the most economical manner and the shortest possible time.

## 1.2 BASIC MEASUREMENTS

Since most engineering surveys are carried out in fairly small areas, the assumption of plane surveying is acceptable. The assumption is that *the Earth's surface is flat*, therefore the lines of gravity (plumb-lines) from all points on the surface are parallel and all measured angles are plane angles. In a triangle of approximately 120 km² on the Earth's surface, the difference between the sum of the spheroidal angles and the plane angles would be about 1″ of arc. The above assumption is therefore quite legitimate, and points located on the irregular surface of the Earth may be projected vertically on to a plane surface in their presentation on a plan (*Figure 1.1*).

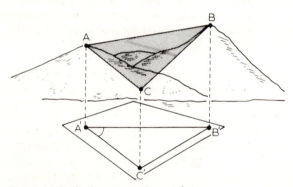

*Figure 1.1* Projection onto a plane surface

An examination of *Figure 1.1* shows clearly the basic surveying measurements needed to locate points *A*, *B* and *C* and to plot them orthogonally as *A′*, *B′* and *C′*. In the first instance the measured *slant distance AB* will fix the position of *B* relative to *A*. However, the *vertical angle* from *A* to *B* will also be required in order to reduce *AB* to its equivalent *horizontal distance A′B′* for the purposes of plotting. Similar measurements will fix *C* relative to *A*, but to fix *C* relative to *B* requires the *horizontal angle BAC* (*B′A′C′*). The *vertical distances* defining the relative elevations of the three points may also be obtained either from the slant distance and vertical angle (trigonometrical levelling), or by direct levelling (Chapter 2) relative to a specific reference datum. The five measurements mentioned above comprise the basis of plane surveying and are illustrated in *Figure 1.2*, i.e. *A′B* is the slant distance, *A′B′* the horizontal distance, *B′B* the vertical distance, *BA′B′* the vertical angle ($\alpha$) and *B′A′C′* the horizontal angle ($\theta$).

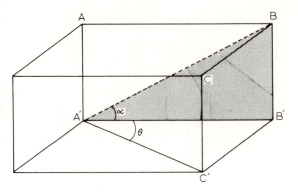

*Figure 1.2* Basic measurements

It can be seen from the above that the only measurements needed in plane surveying are *angles* and *distances*. Nevertheless, the full impact of modern technology has been brought to bear in the acquisition and processing of this simple data. Angles are now easily resolved to single-second accuracy by using optical and electronic theodolites; electromagnetic distance-measuring (EDM) equipment can obtain distances of several kilometers to sub-millimetre precision; lasers and north-seeking gyroscopes are virtually standard equipment for tunnel surveys; orbiting satellites and inertial survey systems (spin-offs from the space programme) are being used for position fixing off-shore as well as on-shore; continued improvement in aerial and terrestrial photo-grammetric equipment and remote sensors makes photogrammetry an invaluable surveying tool; finally, data loggers and computers enable the most sophisticated procedures to be adopted in the processing and automatic plotting of field data.

## 1.3 CONTROL NETWORKS

The establishment of two- or three-dimensional control networks is the most fundamental operation in the surveying of an area whether it is of large or small extent. Their concept can best be illustrated by considering the survey of a relatively small area of land, as shown in *Figure 1.3*.

The processes involved in carrying out the survey can be itemized as follows:

(1)  A careful reconnaissance of the area is first carried out in order to establish the most suitable positions for the survey stations (or control points) *A*, *B*, *C*, *D*, *E* and *F*. The stations should be intervisible and so positioned as to afford easy and accurate measurement of the distances between them. They should form 'well-conditioned' triangles with all angles greater than 45°, and the sides of the triangles should lie close to the topographic detail to be surveyed. Adopting this procedure eliminates the problems of measuring up, over or around obstacles.

The survey stations themselves may be stout wooden pegs driven well down into the ground, with a fine nail in the top accurately depicting the survey position. Alternatively, for longer life, concrete blocks may be set into the ground with some form of fine mark to pin-point the survey position.

(2) The distances between the survey stations are now obtained to the required accuracy. Steel tapes or chains may be laid along the ground to measure the slant lengths, whilst vertical angles may be measured by using hand-held clinometers or

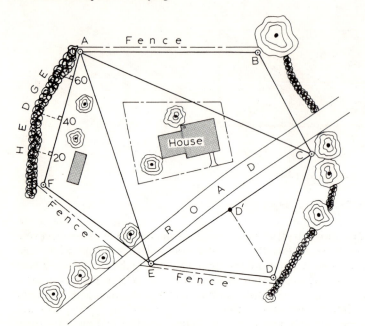

*Figure 1.3* Linear survey

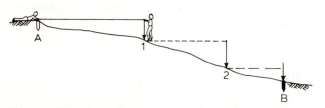

*Figure 1.4* Stepped measurement

Abney levels to reduce the lengths to their horizontal equivalents. Alternatively, the distances may be measured in horizontal steps as shown in *Figure 1.4*. The steps are short enough to prevent sag in the tape, and their end positions at 1, 2 and *B* are fixed with a plumb bob and an additional assistant. The steps are then summed to give the horizontal distance.

Thus by measuring all the distances, the relative positions of the survey stations are located at the intersections of the straight lines and the network possesses shape and scale. The surveyor has thus established in the field a two-dimensional horizontal control network whose nodal points are accurately fixed at known positions.

It must be remembered, however, that all measurements, no matter how carefully carried out, contain error. Thus, as the three sides of a triangle will always plot to give a triangle, regardless of the error in the sides, some form of independent check should be introduced to reveal the presence of error. In this case (*Figure 1.3*) the horizontal distance from *D* to a known position *D'* on the line *EC* is measured. If this distance will not plot correctly within the plotted triangle *CDE*, then error is present in one or all of the sides of the triangle. Similar checks should be introduced throughout the network to prove its reliability.

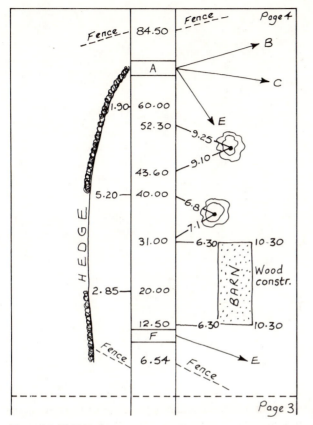

*Figure 1.5* Field book

(3) The proven network can now be used as a reference framework or huge template from which further measurements can be taken to the topographic detail. For instance, in the case of line *FA*, its position may be physically established in the field by aligning a tape or chain between two survey stations. Now, offset measurements taken at right angles to this line at known distances from *F*, say, 20 m, 40 m and 60 m, will locate the position of the hedge. Similar measurements from the remaining lines will locate the positions of the remaining detail.

The method of booking the data for this form of survey is illustrated in *Figure 1.5*. The centre column of the book is regarded as the survey line *FA* with distances along it and offsets to the topographic details drawn in their relative positions, as shown in *Figure 1.3*.

Note the use of oblique offsets to the trees to fix their positions more accurately by intersection, thereby eliminating the error of estimating the right-angle in the other offset measurements.

The network is now plotted to the required scale, the offsets are plotted from the network and the relative positions of all the topographic detail are established to form a plan of the area.

(4) As the aim of this particular survey was the production of a plan, the accuracy of the survey is governed largely by the scale of the plan. For instance, if the scale was, say, one

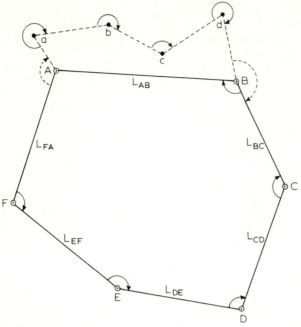

*Figure 1.6*

part in 1000, then a plotting accuracy of 0.1 mm would be equivalent to 100 mm on the ground and it would not be economical or necessary to take the offset measurements to any greater accuracy than this. However, as the network forms the reference base from which the measurements are taken, its position would need to be fixed to a much greater accuracy.

The foregoing comprises the steps necessary to carry out this particular form of survey, which is generally referred to as a *chain* or *linear survey*. It is a process naturally limited to quite small areas, due to the difficulties of measuring with tapes or chains and the rapid accumulation of error involved in the process. For this reason it is a surveying technique not widely used, although it does serve to illustrate in a simple, easy-to-understand manner the basic concepts of *all* surveying.

Had the area been much greater in extent, the distances could have been measured by EDM equipment; such a network is called a *trilateration*. A further examination of *Figure 1.3* shows that the shape of the network could be established by measuring all the horizontal angles, with its scale or size being fixed by the measurement of one side. This network would be a *triangulation*. If all the sides and horizontal angles are measured the network is a *triangulateration*. Finally, if the survey stations are located by measuring the adjacent angles and lengths, as shown in *Figure 1.6* (thereby constituting a polygon *ABCDEF*), then the network is called a *traverse* (see Chapter 4).

These then constitute all the basic methods of establishing a horizontal control network, and they are dealt with, in detail, in Volume 2 of this work.

## 1.4 LOCATING POSITION

The method of locating the position of topographic detail by right-angled offsets from the sides of the control network has been mentioned above (see *Figure 1.3*). However, this method would have errors in establishing the line *FA*, in setting out the right-angle (usually by eye), and in measuring the offset. It would, therefore, be more accurate to locate position directly from the survey stations. The most commonly-used method of doing this is by *polar co-ordinates*, as shown in *Figure 1.7*. A and B are survey stations of known position in a control network, from which the measured horizontal angle *BAP* and the horizontal distance *AP* will fix the position of point *P*. There is no doubt that

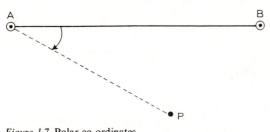

*Figure 1.7* Polar co-ordinates

this is the most popular method of fixing position, particularly since the advent of EDM equipment. Indeed, the method of traversing is a repeated application of this process.

An alternative method is by *intersection* where *P* is fixed by measuring the horizontal angles *BAP* and *ABP*, as shown in *Figure 1.8*. This method forms the basis of triangulation. Similarly, *P* may be fixed by the measurement of horizontal distances *AP* and *BP*, and this forms the basis of the method of trilateration. In both these instances there is no independent check as a position for *P* (not necessarily the correct one) will always be obtained. Thus, at least one additional measurement is required either by combining the angles and distances (triangulateration), by measuring the angle at *P* as a check on the angular intersection, or by producing a trisection from an extra control station.

The final method of position fixing is by *resection*, *Figure 1.9*. This is done by observing the horizontal angles at *P* to at least three control stations of known position. The position of *P* may be obtained by either a graphical or mathematical solution, as illustrated in Volume 2, Chapter 2.

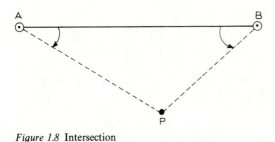

*Figure 1.8* Intersection

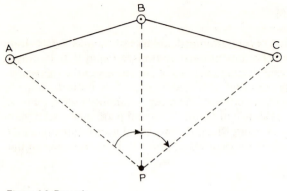

*Figure 1.9* Resection

## 1.5 ERRORS IN MEASUREMENT

It should now be apparent that position-fixing involves simply the measurement of angles and distance. However, no matter how carefully executed, all measurements will contain error, and so the true value of a measurement is never known. It follows from this that if the true value is never known, then the true error can never be known either.

The sources of error fall into three broad categories, namely:

(1) Natural errors caused by variable or adverse weather conditions; refraction; gravity effects, etc.
(2) Instrumental errors caused by imperfect construction and adjustment of the surveying instruments used.
(3) Personal errors caused by the inability of the individual to make exact observations due to the limitations of human sight, touch and hearing.

## 1.6 TYPES OF ERROR

Ignoring mistakes or blunders caused largely by carelessness or total inexperience, errors of observation fall into two categories, (a) systematic errors, and (b) random errors (or variates).

(1) *Systematic errors:* These conform to mathematical and physical laws; thus it is argued that to reduce their effect appropriate corrections can be computed and applied. A simple example of such an error is the expansion of a steel tape due to increase in its temperature above its standard temperature. Thus, if the standard temperature is known (usually 20°C) and the temperature at the time of measurement is measured by thermometer, the expansion can be calculated and the correction applied. Whilst in theory this appears correct, it is doubtful if all systematic error is ever removed from the measuring process. For instance, the thermometer may have an index error, whilst the air temperature measured may not be that of the tape material. Therefore the effect of systematic error can be reduced but never entirely eliminated. Also, as they can be entirely positive or entirely negative and not random in nature, repeated observations will only increase their effect. Systematic errors then are the most difficult to deal with and worthy of very careful consideration prior to, during and after the survey.

(2) *Random errors:* Those variates which remain after mistakes and systematic errors have been minimized. They are beyond the control of the observer, and are caused by the human inability of the observer to make exact measurements for reasons already indicated in *Section 1.5*. These errors are assumed to obey the law of probability and have a normal distribution. It is these errors alone therefore that can be statistically analysed and their effect evaluated.

The main characteristics of random errors as indicated by a typical bell-shaped probability curve are:

(a) Small errors occur more frequently than large ones; that is, they are more probable.
(b) Positive and negative errors of equal magnitude occur with equal frequency; that is, they are equally probable.
(c) Large errors occur infrequently; that is, they are less probable and are more likely to be mistakes or untreated systematic errors.

There is no ideal method of dealing with the random errors which occur in the observation of control networks. The methods used vary from the simple non-mathematical to the more complex methods based on the statistical theory of least squares. Yet regardless of their mathematical rigor the error distribution is frequently at variance with the law of probability and the common-sense logic of the observer. For this reason the greatest care should be given to meticulous observation and field procedures until future research produces more viable methods of error adjustment.

## 1.7 SUMMARY

In the preceding paragraphs an attempt has been made to outline the basic concepts of surveying. Because of their importance these concepts are now summarized as follows:

(1) *Reconnaissance* is the first and most important step in the surveying process. Only after a careful and detailed reconnaissance of the area can the surveyor decide upon the techniques and instrumentation required to complete the work economically and meet the accuracy specifications.
(2) *Control networks* form not only a reference framework for locating the position of topographic detail and setting-out constructions, they may also be used as a base for minor control networks containing a greater number of control stations at shorter distances apart and to a lower order of accuracy, i.e. *a, b, c, d* in *Figure 1.6*. These minor control stations may be more judicially placed for the purpose of locating the topographic detail.

This process of establishing the major control first to the highest order of accuracy, as a framework on which to connect the minor control, which is in turn used as a reference framework for detailing, is known as *working from the whole to the part* and forms the basis of all good surveying procedure.
(3) *Errors* are contained in all measurement procedures and a constant battle must be waged by the surveyor to minimize their effect.

It follows from this that the greater the accuracy specifications the greater the cost of the survey, for it results in more observations, taken with greater care, over a longer period of time, using more precise (and therefore more expensive) equipment. It is for this reason that major control networks contain the minimum number of stations necessary and surveyors adhere to the economic principle of working to an accuracy neither greater than, nor less than, that required.

(4) *Independent checks* should be introduced not only into the field work, but also into the subsequent computation and reduction of field data. In this way, major errors can be quickly recognized and dealt with.

(5) *Commensurate accuracy* is advised in the measuring process; that is, that the angles should be measured to the same degree of accuracy as the distances and *vice versa*. The following rule is advocated by most authorities for guidance—1″ of arc subtends 1 mm at 200 m.

This means that if distance is measured to, say, 1 in 200 000, the angles should be measured to 1″ of arc, and so on.

The author's experience of commensurate accuracy is that distances are twice as strong as angles. For instance, measurement of the three angles in a triangle gives only its shape, whilst measurement of its three lengths gives shape and scale. Thus, whilst the above rule is a useful guide, it is no more than that.

This then constitutes the overall basic approach to the surveying concepts used in the ultimate production of a plan.

# Simple and precise levelling

## 2.1 BASIC CONCEPTS

*Levelling* is the process of obtaining the vertical heights of specific points on the Earth's surface above or below a datum plane.

The basic concept of levelling can be clearly illustrated using a very simple example. Consider *Figure 2.1(a)* in which it is required to find the vertical height of the filing cabinet above the table. This can be easily resolved to a given accuracy by measuring the vertical height $H$, as shown, with a steel tape. However, the problem is more complex if the table and cabinet are in separate rooms (*Figure 2.1(b)*). It may be resolved, by measuring the heights $H_1$ and $H_2$ to the tops of the table and cabinet respectively, assuming that the floor on which they stand is horizontal. That is, the vertical heights are measured relative to a common datum. Such heights are then termed *spot levels*. For instance, if $H_1 = 1$ m and $H_2 = 1.5$ m, then the 'level' of the table is 1 m and of the cabinet 1.5 m, and the difference in levels is 0.5 m, which is the difference in height required. This then is the basic concept of levelling, the process of obtaining vertical heights above or below a given datum.

If one considers levels over the ground surface of a small island such as the UK, one naturally considers using *mean sea level* (MSL) as a common datum. In order to make this common datum easily accessible to all users throughout the UK, permanent marks, whose levels relative to MSL are known, are established throughout the country. These marks, termed *bench marks* (BM), are set up by the *Ordnance Survey* (OS), the national survey organization for the UK.

Thus, to be more specific, the datum for levelling in the UK is known as the *ordnance datum* (OD) and is the mean of continuous sea-level observations taken over a period of six years at a tidal observatory in Newlyn, Cornwall. On many engineering schemes a local datum of purely arbitrary value is used. A permanent mark is fixed in a stable situation and allotted such a value that the lowest level on the site will still possess a positive value. For example, if the local datum is assigned to be 100 m and the vertical depth to the lowest point 30 m, then the level of the lowest point would be 70 m. Had the vertical depth been 115 m, then the level would have been $-15$ m. Minus signs can be misinterpreted, erased or simply left off, thereby resulting in serious mistakes. They should therefore be avoided by using an adequate value for the adopted datum.

As already stated, ordnance datum (OD) is made easily accessible to everyone by the establishment throughout the UK of a network of marks whose heights above OD

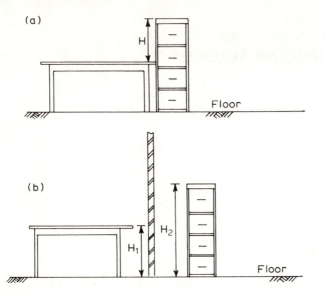

Figure 2.1

have been fixed by the OS. These marks are called *bench marks* (BM), and their positions and values are shown on OS maps at the 1/2500 and 1/1250 scales. The most common type of BM to be found in the UK is the broad-arrow symbol cut into stable, vertical surfaces such as walls, as shown in *Figure 2.2(a)*. Another, less common, type is shown in *Figure 2.2(b)*.

The values of OS BM are quoted and guaranteed only to the nearest 10 mm. Any BM established by persons other than the OS are referred to as *temporary bench marks* (TBM). These may have been tied into the OD or they may have purely arbitrary values.

The majority of engineering work does not require its heights relative to MSL and so the use of an arbitrary datum may be expedient as long as the relative level values are accurate. However, use of OD generally affords the engineer abundant BM for the checking of his levelling circuits and ensures that the site levels are relative to adjoining land which may have been contoured by the OS.

## 2.2 EQUIPMENT

The equipment used in the levelling process comprises optical levels and graduated staffs. Basically an optical level consists of a telescope fitted with a spirit bubble, or automatic compensator to ensure long horizontal sights on to the graduated, vertically-held staff, as shown in *Figure 2.3*. Thus, by reference to *Figure 2.10* it can be seen that the horizontal plane of sight swept out by the level, is itself another datum plane from which vertical distances are measured.

### 2.2.1 Staffs

The levelling staff is of either wood or metal and calibrated in metres and decimals thereof. The British Standards Institution (BSI) have adopted the E-pattern staff, the

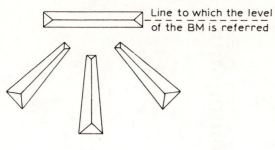

(a)

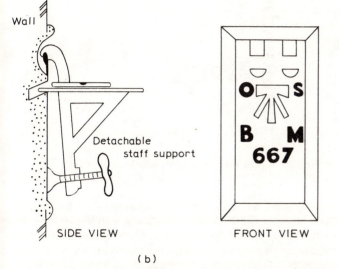

(b)

*Figure 2.2* (a) Cut arrow type bench mark. (b) Flush bracket bench mark

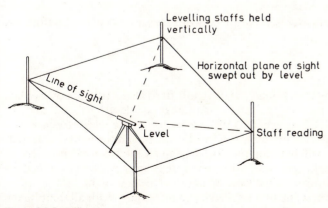

*Figure 2.3* Levelling

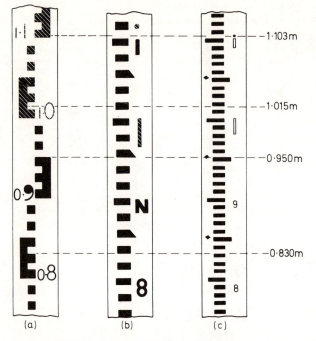

*Figure 2.4* Metric staffs. (*a*) British Standard E-pattern interval. (*b*) Sopwith 10-mm interval. (*c*) 5-mm interval

smallest division of which is 10 mm, but it is read by estimation to the nearest millimetre, e.g. 1.103 m. Also used is the Sopwith metric staff which differs only in the pattern of the graduations. Other metric staffs graduated to 5 mm are available. The author's experience of all three types shows that many more reading errors occur with the 5-mm type, and that the 10-mm E-pattern type is preferred (British Standard metric staff) (*Figure 2.4*).

## 2.2.2 Levels

Although there is a variety of makes, there are only three basic types of level, as follows:

### (1) Dumpy level

The telescope of the dumpy level (*Figure 2.5*) is rigidly fixed to the tribrach or levelling plate. Movement of the footscrews against the trivet stage enables the tribrach to be set horizontal. A sensitive bubble attached to the side or top of the telescope ensures a horizontal line of sight when the instrument is in adjustment and the bubble is at mid-run. A modern instrument will have internal focusing and a Ramsden eyepiece, giving an inverted image. The diaphragm, in addition to the cross-hairs, will carry stadia hairs for three-wire levelling or for roughly ascertaining the length of sight.

Once the dumpy has been set up and levelled it should remain so for all sights taken at that point. It is thus considered ideal for contouring or any work involving many

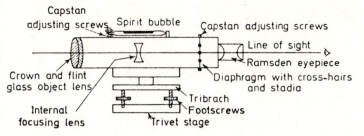

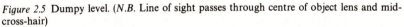

*Figure 2.5* Dumpy level. (*N.B.* Line of sight passes through centre of object lens and mid-cross-hair)

sights taken radially at each set-up. Undue activity around the instrument, tripod settlement, vibration on site, etc. will obviously disturb the verticality of the instrument and require a re-levelling of the bubble. Several such re-levellings will eventually alter the height of the line of sight and result in error. It is thus less accurate than the tilting level with which each line of sight is levelled.

## (2) Tilting level

With this instrument (*Figure 2.6*) the telescope is not rigidly fixed to the tribrach but is pivoted, usually about its centre. A small circular spirit level on the tribrach allows approximate levelling of the instrument. Accurate levelling of the telescope for each line of sight is by means of a tilting screw and sensitive longitudinal bubble. The tilting level may be set up more quickly than the dumpy level for a set-up involving just two or three sights, and these readings will generally be more accurate; it is thus better for fly or sectional levelling.

A *reversible level* is a tilting level which may be rotated about its line of sight to give bubble left and bubble right readings, the average being free of collimation error. It also permits an easy method of adjustment.

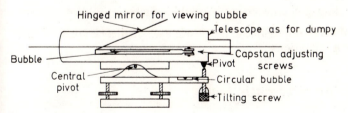

*Figure 2.6* Tilting level

## (3) Automatic or self-aligning level

This instrument (*Figure 2.7(a)*) is like the dumpy in that the telescope is rigidly fixed to the tribrach. A small circular spirit bubble allows approximate levelling with the final accurate levelling being completed automatically by a 'stabilizer' inside the telescope.

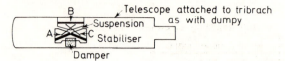

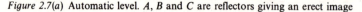

Figure 2.7(a) Automatic level. A, B and C are reflectors giving an erect image

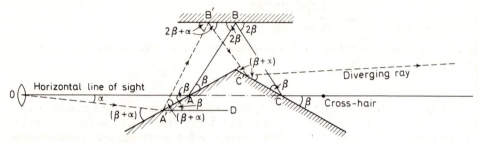

Figure 2.7(b) Reflectors remaining fixed

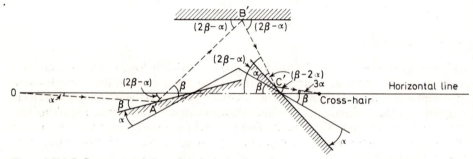

Figure 2.7(c) Reflectors A and C moving clockwise through α

The advantages of this level over the previous two are:

(a) Much simpler to use, as it gives an erect image.
(b) Very rapid operation giving greater economy.
(c) No chance of bubble setting error.
(d) No chance of reading staff without setting bubble first.

One disadvantage is that it is impossible to use on a site where there is great vibration from, say, wind or piling operations.

### 2.2.3 Principle of the automatic level

In the auto-level, the stabilizer ensures that the entering ray of light passes through the cross-hairs, even though the telescope is slightly tilted. In practice:

(1) The entering ray is horizontal.
(2) The telescope, and thus the fixed prism at B, are tilted due to initial approximate levelling.
(3) The freely suspended surfaces at A and C remain at the same angle to the horizontal plane.

In order to illustrate the principle it is imagined that:

(4) The entering ray enters at an angle equal to the tilt of the telescope.
(5) The telescope and prism $B$ remain horizontal.
(6) The surfaces at $A$ and $C$ tilt through the same angle as the entering ray.

*Figure 2.7(b)* shows the ray of light $OA$ entering the telescope when it is perfectly horizontal and being deflected at $A$, $B$ and $C$ to emerge horizontally through the cross-hairs.

Assume now that the telescope is only approximately levelled and ray $OA'$ enters at an angle $\alpha$, the reflectors at $A, B, C$ remaining *fixed*. Then from *Figure 2.7(b)*:

Angle of incidence at $A' = (\beta + \alpha) =$ angle of reflection at $A'$
Angle $B'A'D = (2\beta + \alpha) =$ angle of incidence at $B'$
$= $ angle of reflection $B'$
$\therefore$ Angle of incidence at $C' = (\beta + \alpha) =$ angle of reflection at $C'$

Thus the ray of light will *diverge* from the horizontal by $\alpha$ and fail to meet at the cross-hair.

In *Figure 2.7(c)*, if the telescope is tilted through $\alpha$, substitute the situation outlined in (4), (5) and (6).

Angle of incidence at $A' = \beta =$ angle of reflection at $A'$
from which:
Angle of incidence at $B' = (2\beta - \alpha) =$ angle of reflection at $B'$.

Inspection of *Figure 2.7(c)* shows that the angle of incidence at $C'$ is $(\beta - 2\alpha)$, equal to the angle of reflection. Thus the ray of light now converges on the horizontal by $3\alpha$, the spacing of the stabilizer relative to the reticule ensuring that it passes through the cross-hairs.

## 2.3 INSTRUMENT ADJUSTMENT

In order for equipment to give the best possible results it should be frequently tested and adjusted. Such adjustments are called *permanent adjustments*, and these are described below.

### 2.3.1 Dumpy level

#### (1) Plate bubble test

To ensure that the axis of the bubble is perpendicular to the vertical axis of the instrument.

*Test*

(a) Set bubble parallel to two footscrews. Centralize it, turn it through 90° in the horizontal plane until over the third footscrew, and re-centralize the bubble using this footscrew only. Repeat until bubble remains central for both positions.
(b) Now re-set bubble parallel to two footscrews and very carefully centralize it.

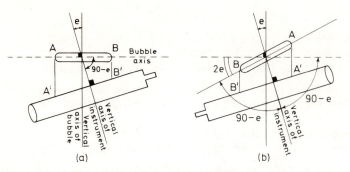

*Figure 2.8* (*a*) With bubble horizontalized. (*b*) When turned through 180°

Assuming that the bubble axis is not at right-angles to the vertical axis but is in error by an amount *e*, then the situation will be as in *Figure 2.8(a)*.

(c) Turn the bubble about the instrument axis through 180° in the horizontal plane; the amount by which it moves off centre is equal to *twice* the instrument error (2*e*) (*Figure 2.8(b)*).

## Adjustment

Bring the bubble half-way back to a central position using the two footscrews. This makes the vertical axis of instrument move through *e* and coincide with the true vertical, thus the level is now usable if no adjusting tools are immediately available. However, the bubble will still be inclined at *e* to the horizontal, so bring the bubble to the centre of its run by raising or lowering one end of the bubble by means of the capstan adjusting screws.

## (2)  Collimation error

To ensure that the line of sight is perpendicular to the vertical axis when the instrument is truly levelled.

## Two-peg test

(a) Set up the instrument mid-way between two pegs, *A* and *B*, about 60 m apart, giving staff readings of, say, 3 m at *A* and 2 m at *B*, as in *Figure 2.9(a)*. Assuming that the line of sight is inclined down from the horizontal by *e*; as this error is proportional to the length of sight and the sights are equal, then the errors at *A* and *B* will be equal and cancel each other out. *Thus the information obtained here is simply that 'A* is lower than *B* by 1 m'. (Note that *e* is called the *collimation error*.)

(b) Now move the instrument to *D*, in line with *A* and *B* and, say, 15 m from *B* (*Figure 2.9(b)*). As the sights to *A* and *B* are no longer equal, the error will be greater at *A* than at *B*. Assume the readings at *A* and *B* to be 4 m and 3.5 m respectively; thus *A* appears lower than *B* by 0.5 m. Knowing that this is not the true difference in level, it is immediately apparent that collimation error is present. If one now constructs a horizontal line from the reading 3.5 m at *B*, it will give a reading of 4.5 m at *A*, as *A* is

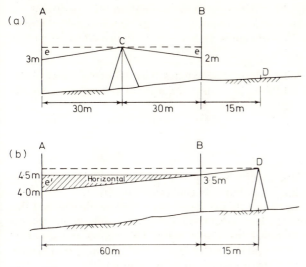

Figure 2.9

truly lower than *B* by 1 m. Thus it can be seen that the collimation error *e'* is 0.5 m in 60 m pointing *down*. Therefore the amount of error from the instrument position at *D* is $(0.5/60) \times 75 = 0.625$ m.

True reading on *A* from $D = 4 + 0.625 = 4.625$ m.

A rule for finding the direction of collimation error is 'If the *false* difference in level is *greater* than the *true* difference then the direction is *up*, and *vice versa*'.

## Adjustment

The instrument is set to read 4.625 m at *A* by raising or lowering (in this case lowering) the cross-hairs by use of the capstan adjusting screws. *Note*. The two-peg test frequently forms an exam question and students are recommended to use the above method, thus eliminating the need to remember formulae. In using the method, only the second sketch need be drawn and with practice, this too can be dispensed with. Regardless of whether or not the collimation error is up or down, it can always be drawn down, the figures will clearly indicate its direction later. A further example will now be worked to illustrate this.

## Example on two-peg test

Assuming the same distances as in the previous example, the readings at *A* and *B* with the level mid-way are 2.85 m and 1.55 m respectively. With the level at *D*, the readings were $A = 3.75$ m and $B = 1.85$ m. Calculate the amount and direction of collimation error, also the true reading at *A* from *D*.

(a) From mid-readings, *A* is *lower* than *B* by 1.3 m (*true*).
(b) From *D* readings, *A* is *lower* than *B* by 1.9 m (*false*), thus collimation error is present.

(c)  Difference between (a) and (b) is the amount of collimation error = 0.6 m.
(d)  From the above rule the direction is *up*. Thus line of sight is too *high* by 0.6 m in 60 m. Collimation error in 75 m (distance $AD$) = 0.75 m. True reading on $A$ from $D = 3.75 - 0.75 = 3$ m.

As shown the question is answered without any sketches. The difference between the 'true' and 'false' readings gives the amount of collimation error. The 'rule' provides the direction.

### 2.3.2 Tilting level

(1)  The first test as described for the dumpy level should, theoretically, be carried out on the small circular bubble of this level, but it is generally ignored.
(2)  The two-peg test is carried out in exactly the same way as for the dumpy but the method of adjustment is different. The level is set to read the true value on the staff by using the tilting screw to raise or lower the line of sight. This causes the sensitive longitudinal bubble to move off centre and this is then corrected by turning the capstan screws. (Note the difference in adjustment compared with dumpy.)
(3)  A third adjustment is to ensure that the tilting screw is in the centre of its run when the main spirit level is central and the vertical axis is truly vertical. This is carried out as follows:

(a)  Set main spirit level parallel to two footscrews and centralize it.
(b)  Turn through 180° and remove any movement off centre, half by the footscrews and half by the capstan adjusting screws.
(c)  Repeat the procedure over the third footscrew until the bubble remains central for any position of the telescope.
(d)  Now slacken off the locking nut of the tilting screw which can then be moved up or down its sleeve until the mid-run position is reached; tighten locking nut. The mid-run position is usually indicated by the coincidence of graduation marks.

This test is carried out only when necessary and on such occasions it should be done prior to the two-peg test.

### 2.3.3 Automatic level

The first test as described for the dumpy should be carried out on the circular bubble of the automatic level, half the error being eliminated by the footscrews and half by the capstan screws of the bubble.

In this case it is very important that the bubble be kept in adjustment otherwise the stabilizer may stick or give inaccurate readings. The stabilizer gives the most accurate results near the centre of its movement, thus the circular bubble needs to be in adjustment to keep the movement of the stabilizer to a minimum. In precise levelling the bubble must be very accurately centred.

The plane of the pendulum swing of the freely-suspended surfaces should be parallel to the line of sight, otherwise a slight error in compensation occurs. Thus, if the circular bubble is in error transversely and the telescope is always pointed in the same direction at each set up, the backsight (BS) will always contain a certain error and the foresight (FS) the same error but of opposite sign. Assuming BS error is $+e$ and FS error $-e$, as the FS is always subtracted from the BS the error becomes $+e - (-e) = 2e$. To avoid this accumulation of systematic error, the telescope should be levelled pointing to the BS and the BS and FS readings taken. It is now re-levelled pointing to the FS and the readings repeated. The mean of the results is free from stabilizer error. This procedure is necessary only for precise work.

The two-peg test is carried out as described and the line of sight adjusted by moving the cross-hair up or down, in exactly the same way as for the dumpy level.

## 2.4 PRINCIPLE OF LEVELLING

The instrument is set up at $A$ (as in *Figure 2.10*) from which point a horizontal line of sight is possible to the TBM at $1A$. The first sight to be taken is to the staff held vertically on the TBM and this is called a *backsight* (BS) the value of which (1.5 m) would be entered in the appropriate column of a levelling book. Sights to points $2A$ and $3A$ where further levels relative to the TBM are required, are called *intermediate sights* (IS), again entered in the appropriate column of the levelling book. The final sight from this instrument set up at $4A$ and is called the *foresight* (FS). It can be seen from the *Figure*, that this is as far as one can go with this sight. If for instance the staff had been placed at $X$, it would not have been visible and would have had to be moved down the slope, towards the instrument at $A$, until it was visible. As foresight $4A$ is as far as one can see from $A$, then it is also called the *change point* (CP), signifying a change of instrument position to $B$. To achieve continuity in the levelling the staff must remain at *exactly the same point* $4A$ although it must be turned to face the instrument at $B$. It now becomes the BS for the new instrument set up and the whole procedure is repeated as before.

Thus, one must remember that all levelling commences on a BS and finishes on a FS with as many IS in between as are required; and that change points are always FS/BS. Also, it must be closed back into a known BM to ascertain the misclosure error.

### 2.4.1 Reduction of levels

From *Figure 2.10*, realizing that the line of sight from the instrument at $A$ is truly horizontal, it can be seen that the higher reading of 2.5 at point $2A$ indicates that the point is lower than the TBM by 1.0, giving $2A$ a level therefore of 59.5. This can be written as follows:

$1.5 - 2.5 = -1.0$, indicating a *fall* of 1.0 from $1A$ to $2A$
Level of $2A = 60.5 - 1.0 = 59.5$

Similarly between $2A$ and $3A$, the higher reading on $3A$ shows it is 1.5 below $2A$, thus:

$2.5 - 4.0 = -1.5$ (fall from $2A$ to $3A$)
Level of $3A$ = level of $2A - 1.5 = 58.0$

Finally the *lower* reading on $4A$ shows it to be *higher* than $3A$ by 2.0, thus:

$4.0 - 2.0 = +2.0$, indicating a *rise* from $3A$ to $4A$
Level of $4A$ = level of $3A + 2.0 = 60.0$

Now, knowing the *reduced level* (RL) of $4A$, i.e. 60.0, the process can be repeated for the new instrument position at $B$. This method of reduction is called the *rise-and-fall (R-and-F) method*.

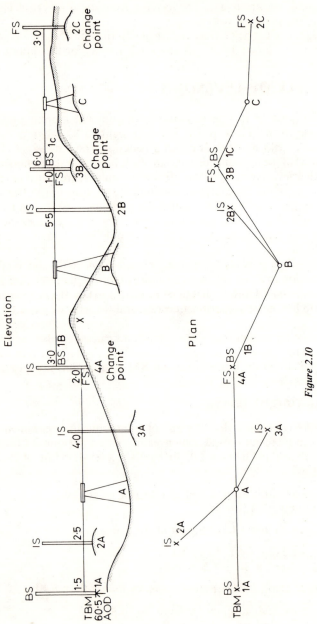

Elevation

Plan

*Figure 2.10*

## 2.4.2  Methods of booking

### (1)  Rise-and-fall

| BS | IS | FS | Rise | Fall | RL | Distance | Remarks | |
|---|---|---|---|---|---|---|---|---|
| 1.5 | | | | | 60.5 | 0 | TBM (60.5) | 1A |
| | 2.5 | | | 1.0 | 59.5 | 30 | | 2A |
| | 4.0 | | | 1.5 | 58.0 | 50 | | 3A |
| 3.0 | | 2.0 | 2.0 | | 60.0 | 70 | CP | 4A (1B) |
| | 5.5 | | | 2.5 | 57.5 | 95 | | 2B |
| 6.0 | | 1.0 | 4.5 | | 62.0 | 120 | CP | 3B (1C) |
| | | 3.0 | 3.0 | | 65.0 | 160 | TBM (65.1) | 2C |
| 10.5 | | 6.0 | 9.5 | 5.0 | 65.0 | | checks | |
| 6.0 | | | | 5.0 | 60.5 | | misclosure | 0.1 |
| 4.5 | | | 4.5 | | 4.5 | | *correct* | |

The above extract of booking is largely self-explanatory. Students should note:

(a) Each reading is booked on a separate line except for the BS and FS at change points. The BS is booked on the same line as the FS because it refers to the same point. As each line refers to a specific point it should be noted in the remarks column.

(b) Each reading is substracted from the previous one, i.e. 2A from 1A, then 3A from 2A, 4A from 3A and stop; the procedure re-commencing for the next instrument station, 2B from 1B and so on.

(c) Three very important checks must be applied to the above reductions, namely:

the sum of BS − the sum of FS = sum of rises − sum of falls
= last reduced level − first reduced level

These checks are shown in the above Table. It should be emphasized that they are nothing more than checks on the arithmetic of reducing the levelling results, they are in no way indicative of the accuracy of the fieldwork.

(d) It follows from the above that the first two checks should be carried out and verified before working out the reduced levels (RL).

(e) Closing error = 0.1, and can be assessed only by connecting the levelling into a BM of known and proven value or connecting back into the starting BM.

### (2)  Height of collimation

This is the name given to an alternative method of booking. The reduced levels are found simply by subtracting the staff readings from the reduced level of the line of sight (plane of collimation). In *Figure 2.10*, for instance, the *height of the plane of collimation* (HPC) at $A$ is obviously $(60.5 + 1.5) = 62.0$; now 2A is 2.5 below this plane, thus its level must be $(62.0 − 2.5) = 59.5$; similarly for 3A and 4A to give 58.0 and 60.0 respectively. Now the procedure is repeated for $B$. The tabulated form shows how simple this process is:

| BS | IS | FS | HPC | RL | Remarks | |
|----|-----|-----|------|------|-------------|-----|
| 1.5 | | | 62.0 | 60.5 | TBM (60.5) | 1A |
| | 2.5 | | | 59.5 | | 2A |
| | 4.0 | | | 58.0 | | 3A |
| 3.0 | | 2.0 | 63.0 | 60.0 | change pt | 4A (1B) |
| | 5.5 | | | 57.5 | | 2B |
| 6.0 | | 1.0 | 68.0 | 62.0 | change pt | 3B (1C) |
| | | 3.0 | | 65.0 | TBM (65.1) | 2C |
| 10.5 | 12.0 | 6.0 | | 65.0 | checks | |
| 6.0 | | | | 60.5 | misclosure 0.1 | |
| 4.5 | | | | 4.5 | *correct* | |

Thus it can be seen that:

(a)  BS is added to RL to give HPC, i.e., 1.5 + 60.5 = 62.0.
(b)  Remaining staff readings are *subtracted* from HPC to give the RL.
(c)  Procedure repeated for next instrument set up at B, i.e., 3.0 + 60.0 = 63.0.
(d)  Two checks same as R-and-F method, i.e.:

sum of BS − sum of FS = last RL − first RL.

(e)  The above two checks are not complete; for instance, if when taking 2.5 from 62 to get RL of 59.5, one wrote it as 69.5, this error of 10 would remain undetected. Thus the *intermediate* sights are *not* checked by those procedures in (d) above and the following cumbersome check must be carried out:

sum of all the RL except the first = (sum of each HPC multiplied by the number of IS or FS taken from it) − (sum of IS and FS).

e.g.   $362.0 = [(62.0 \times 3.0) + (63.0 \times 2.0) + (68.0 \times 1.0)]$
$- [12.0 + 6.0] = 362.0$

### 2.4.3 Inverted sights

*Figure 2.11* shows inverted sights at B, C and D to the underside of a structure. It is obvious from the drawing that the levels of these points are obtained by simply adding the staff readings to the HPC to give B = 65.0, C = 63.0 and D = 65.0; E is obtained in the usual way and equals 59.5. However, the problem of inverted sights is completely eliminated if one simply treats them as *negative* quantities and proceeds in the usual way:

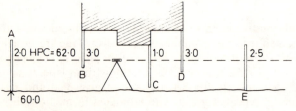

*Figure 2.11* Inverted sights

| BS | IS | FS | Rise | Fall | HPC | RL | Remarks | |
|----|----|----|------|------|-----|----|---------|----|
| 2.0 | | | | | 62.0 | 60.0 | TBM | *A* |
| | −3.0 | | 5.0 | | | 65.0 | | *B* |
| | −1.0 | | | 2.0 | | 63.0 | | *C* |
| | −3.0 | | 2.0 | | | 65.0 | | *D* |
| | | 2.5 | | 5.5 | | 59.5 | TBM | *E* (59.55) |
| 2.0 | −7.0 | 2.5 | 7.0 | 7.5 | | 60.0 | checks | |
| | | 2.0 | | 7.0 | | 59.5 | misclosure 0.05 | |
| | | 0.5 | 0.5 | 0.5 | | 0.5 | *correct* | |

R-and-F method

$$2.0 - (-3.0) = +5.0 = \text{Rise}$$
$$-3.0 - (-1.0) = -2.0 = \text{Fall}$$
$$-1.0 - (-3.0) = +2.0 = \text{Rise}$$
$$-3.0 - \phantom{(}2.5\phantom{)} = -5.5 = \text{Fall}$$

HPC method

$$62.0 - (-3.0) = 65.0$$
$$62.0 - (-1.0) = 63.0$$
$$62.0 - (-3.0) = 65.0$$
$$62.0 - (+2.5) = 59.5.$$

In the checks inverted sights are treated as negative quantities; e.g. check for IS in HPC method gives:

$$252.5 = (62.0 \times 4.0) - (-7.0 + 2.5)$$
$$= 248.0 - (-4.5) = 248.0 + 4.5 \doteq 252.5$$

## 2.4.4 Comparison of methods

In the author's opinion the R-and-F method should be used at all times because of the very simple but complete arithmetical checks involved. The R-and-F columns also give a visual indication of the topography. Although the HPC method involves slightly less arithmetic, particularly where there are numerous IS, as in grid levelling, its one great failing is the cumbersome IS check. However, it is useful when setting out levels as described in *Section 2.6.1(1)*.

It is significant that the report 'Survey Standards, Setting Out and Earthworks Measurement' produced in 1982 by a joint working party of the Institutions of Highway and Civil Engineers, recommends the use of the R-and-F method of booking.

## 2.5 SOURCES OF ERROR

(1) The main source of error is the residual collimation error of the instrument. From the two-peg test it should be apparent that this error would be eliminated by equalizing the lengths of the BS and FS. This could be done quite adequately by using the stadia hairs of the level, as in tacheometry, although in simple levelling balancing of the lengths of sights is not always possible. Theoretically equalization of the sums of the lengths of BS and FS will eliminate collimation error, but variations in focusing may affect this.

(2) Staff not held vertical; eliminated by fitting a spirit bubble to the staff, or by swaying the staff backwards and forwards in the direction of the level until a minimum reading is obtained.

(3) Error in reading staff; minimized by reducing the length of sight so that readings are easily defined.

(4) Mistake in reading staff, such as reading '6' for '9'; have the booker read figures back and check them. May also be used to reduce booking error.

(5) Staff moving off position at CP when turned to face the new instrument setting; use a levelling plate on soft ground, and clearly mark the CP on hard ground.

(6) Instrument settlement; set up on firm ground, dig tripod legs well into the ground and avoid excessive movement about instrument.

(7) Errors due to refraction from warm layers of air at ground level; keep readings at least 1 m above ground.

(8) Errors due to staff not being fully extended. This may be due either to carelessness on the staff-holder's part or to wear on the joints or retaining spring.

(9) Finally, it is important to eliminate parallax in the instrument by bringing the cross-hair into sharp focus using the eye-piece focusing screw.

Students should also note the errors eliminated by the use of automatic levels. The procedure should always be carried out before commencing the levelling.

### 2.5.1 Accuracy

The error in levelling can be found only by closing the circuit back to its starting point or by tying in to another established BM. This latter method should be regarded with caution because the BM themselves contain error, e.g. in levelling from one OS BM to another, one may get a final RL agreeing with that of the BM; indicating no error in the levelling. However, the OS will only guarantee the values of adjacent BM to within 10 mm. Generally, the error $E$ should not exceed $12(K)^{1/2}$ mm, although in hilly terrain with frequent short sights a more realistic value may be twice the above amount, i.e. $E = 24(K)^{1/2}$ mm ($K$ is the distance levelled in kilometres).

## 2.6 CONTOURING

A simple definition of a *contour* is that it is a line joining all points of equal level. Thus contour lines on a plan illustrate the conformation of the ground. For instance, when contour lines are close together they represent steeply-sloping ground, and *vice versa*. Contours are used by the engineer in a variety of ways, as the following sample shows:

(1) In the computation of volumes.
(2) In the construction of lines of constant gradient.
(3) In delineating the limits of construction work. For example, the points of intersection, of the contours (strike lines) of a constructed or proposed slope, with ground contours of equal elevation, when joined up show the limits or slope stake positions of the construction.
(4) In the delineation and measurement of drainage areas.

The constant vertical distance between contours is termed the *contour interval*. The appropriate interval to adopt in any particular instance is dependent upon:

(1) *Cost:* The smaller the interval adopted the greater the amount of work involved, resulting therefore in higher costs.
(2) *Purpose and extent of the survey:* Where a plan is required for detailed design and/or the measurement of earthworks, the interval may be as low as 0.5 m or up to 2 m on larger sites. For general topographic mapping the interval may range from 5 to 20 m depending upon the scale of the plan and the nature of the terrain.

## 2.6.1 Methods of contouring

For general contouring work, vertical-staff tacheometry is probably the most popular method (Chapter 5). However, when greater accuracy is required a level and staff may be used in the following procedures:

(1) *Direct contouring:* In this method the actual contour is pegged out on the ground and its planimetric position surveyed. A back sight is taken to an appropriate TBM and the HPC of the instrument obtained, say 34.800 m OD. Then a reading of 0.800 m on the staff would indicate that the foot of the staff was at the 34 m level. In this way the 34-m contour could be located and pegged out at regular intervals over the terrain. Similarly a reading of 1.800 m gives the 33-m contour, and so on. Contours are established one at a time and their position surveyed using an appropriate technique. As the accuracy of the contour depends not only upon the accuracy of levelling but also upon the accuracy of its position, this may be the controlling factor in the method of survey adopted to fix its position. It may be done by chain surveying, plane tabling or by offsets, polars or intersection from traverse lines.

This technique is usually adopted on sites where the construction is related to a specific contour line, such as excavation work up to a given contour.

(2) *Indirect contouring:* This involves establishing a grid over the area and obtaining levels at the corners of the grid squares. The grid interval will be dependent upon the rugosity of the terrain and the purpose for which the information is required. This latter factor may also control the accuracy to which the grid is established. For instance if the grid is also to be used for dimensional control, it may need to be established to a very high degree of accuracy. Linear interpolation between the levels, on the assumption of uniform slope between them is then used to locate the contours. Where the work in question is for route alignment, levels are taken at regular intervals each side of the route centre-line, along lines normal to the centre-line. The levels are then interpolated for contours and also used direct in computing the areas of cross-sections for earthwork quantities.

## Worked examples

*Example 2.1.* The positions of the pegs which need to be set out for the construction of a sloping concrete slab are shown in *Figure 2.12*. Because of site obstructions the tilting

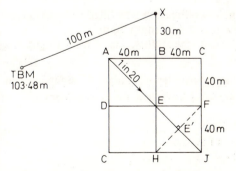

*Figure 2.12 (N.B. Dotted line HF and station E' are not part of the question)*

level which is used to set the pegs at their correct levels can only be set up at station $X$ which is 100 m from the TBM. The reduced level of peg $A$ is to be 100 m and the slab is to have a uniform diagonal slope from $A$ towards $J$ of 1 in 20 downwards.

To ensure accuracy in setting out the levels it was decided to adjust the instrument before using it, but it was found that the correct adjusting tools were missing from the instrument case. A test was therefore carried out to determine the magnitude of any collimation error that may have been present in the level, and this error was found to be 0.04 m per 100 m downwards.

Assuming that the backsight reading from station $X$ to a staff held on the TBM was 1.46 m, determine to the nearest 0.01 m the staff readings which should be obtained on the pegs at $A$, $F$ and $H$, in order that they may be set to correct levels.

Describe fully the procedure that should be adopted in the determination of the collimation error of the tilting level.                                              (ICE)

The simplest approach to this question is to work out the true readings at $A$, $F$ and $H$ and then adjust them for collimation error. Allowing for collimation error the true reading on TBM $= 1.46 + 0.04 = 1.50$ m

$$\text{HPC} = 103.48 + 1.50 = 104.98 \text{ m}$$

True reading on $A$ to give a level of 100 m $= 4.98$ m
Dist $AX = 50$ m $(\Delta AXB = 3, 4, 5)$
$\therefore$  Collimation error $= 0.02$ m per 50 m
Allowing for this error, actual reading at $A = 4.98 - 0.02 = 4.96$ m
Now referring to *Figure 2.12*, line $HF$ through $E'$ will be a strike line
$\therefore$  $H$ and $F$ have the same level as $E'$
Dist $AE' = (60^2 + 60^2)^{1/2} = 84.85$ m
Fall from $A$ to $E' = 84.85 \div 20 = 4.24$ m
$\therefore$  Level at $E' =$ level at $F$ and $H = 100 - 4.24 = 95.76$ m
Thus true staff readings at $F$ and $H = 104.98 - 95.76 = 9.22$ m
Dist $XF = (70^2 + 40^2)^{1/2} = 80.62$ m
Collimation error $\approx 0.03$ m
Actual reading at $F = 9.22 - 0.03 = 9.19$ m
Dist $XH = 110$ m, collimation error $\approx 0.04$ m
Actual reading at $H = 9.22 - 0.04 = 9.18$ m.

*Example 2.2.* The following readings were observed with a level: 1.143 (BM 112.28), 1.765, 2.566, 3.820 CP 1.390, 2.262, 0.664, 0.433 CP 3.722, 2.886, 1.618, 0.616 TBM.

(1) Reduce the levels by the R-and-F method.
(2) Calculate the level of the TBM if the line of collimation was tilted upwards at an angle of 6′ and each BS length was 100 m and FS length 30 m.
(3) Calculate the level of the TBM if in all cases the staff was held not upright but leaning backwards at 5° to the vertical.                                              (LU)

(1) The answer here relies on knowing once again that levelling always commences on a BS and ends on a FS, and that CP are always FS/BS (see Table opposite).
(2) Due to collimation error
the BS readings are too great by 100 tan 6′
the FS readings are too great by  30 tan 6′
*net* error on BS too great by       70 tan 6′

The student should note that the IS are unnecessary in calculating the value of the

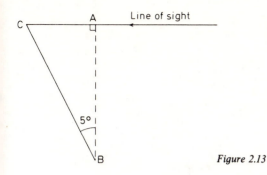

*Figure 2.13*

TBM; he can prove it for himself by simply covering up the IS column and calculating the value of TBM using BS and FS only.

There are three instrument set-ups, therefore the *total net error* on BS = $3 \times 70 \tan 6' = 0.366$ m (too great).

Level of TBM = $113.666 - 0.366 = 113.300$ m

(3)  From *Figure 2.13* it is seen that the true reading $AB$ = actual reading $CB \times \cos 5°$. Thus each BS and FS needs to be corrected by multiplying it by $\cos 5°$; however, this would be the same as multiplying the $\Sigma BS$ and $\Sigma FS$ by $\cos 5°$, and as one subtracts BS from FS to get the difference, then

*True* difference in level = actual difference $\times \cos 5°$
$$= 1.386 \cos 5° = 1.381 \text{ m}$$
Level of TBM $\qquad = 112.28 + 1.381 = 113.661$ m

| BS | IS | FS | Rise | Fall | RL | Remarks |
|---|---|---|---|---|---|---|
| 1.143 | | | | | 112.280 | BM |
| | 1.765 | | | 0.622 | 111.658 | |
| | 2.566 | | | 0.801 | 110.857 | |
| 1.390 | | 3.820 | | 1.254 | 109.603 | |
| | 2.262 | | | 0.872 | 108.731 | |
| | 0.664 | | 1.598 | | 110.329 | |
| 3.722 | | 0.433 | 0.231 | | 110.560 | |
| | 2.886 | | 0.836 | | 111.396 | |
| | 1.618 | | 1.268 | | 112.664 | |
| | | 0.616 | 1.002 | | 113.666 | TBM |
| 6.255 | | 4.869 | 4.935 | 3.549 | 113.666 | |
| 4.869 | | | | 3.549 | 112.280 | |
| 1.386 | | | 1.386 | | 1.386 | checks |

*Example 2.3.* One carriageway of a motorway running due N is 8 m wide between kerbs (*Figure 2.14*) and the following surface levels were taken along a section of it, the chainage increasing from S to N. A concrete bridge 12 m in width and having a horizontal soffit, carries a minor road across the motorway from SW to NE, the centre-line of the minor road passing over that of the motorway carriageway at a chainage of 1550 m.

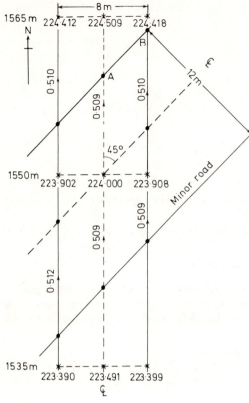

Figure 2.14

| BS | IS | FS | Chainage (m) | Location |
|---|---|---|---|---|
| 1.591 | | | 1535 | west channel |
| | 1.490 | | 1535 | crown |
| | 1.582 | | 1535 | east channel |
| | −4.566 | | | bridge soffit* |
| | 1.079 | | 1550 | west channel |
| | 0.981 | | 1550 | crown |
| | 1.073 | | 1550 | east channel |
| 2.256 | | 0.844 | | CP |
| | 1.981 | | 1565 | west channel |
| | 1.884 | | 1565 | crown |
| | | 1.975 | 1565 | east channel |

* Staff inverted.

Taking crown (i.e. centre-line) level of the motorway carriageway at 1550 m chainage to be 224.000 m

(a) Reduce the above set of levels and apply the usual arithmetical checks.

(b) Assuming the motorway surface to consist of planes, determine the minimum vertical clearance between surface and the bridge soffit. (LU)

The HPC method of booking is used because of the numerous IS.

| BS | IS | FS | HPC | RL | Remarks |
|---|---|---|---|---|---|
| 1.591 | | | | 223.390 | 1535 W channel |
| | 1.490 | | | 223.491 | 1535 crown |
| | 1.582 | | | 223.399 | 1535 E channel |
| | −4.566 | | | 229.547 | bridge soffit |
| | 1.079 | | | 223.902 | 1550 W channel |
| | 0.981 | | 224.981* | 224.000 | 1550 crown |
| | 1.073 | | | 223.908 | 1550 E channel |
| 2.256 | | 0.844 | 226.393 | 224.137 | CP |
| | 1.981 | | | 224.412 | 1565 W channel |
| | 1.884 | | | 224.509 | 1565 crown |
| | | 1.975 | | 224.418 | 1565 E channel |
| 3.847 | 5.504 | 2.819 | | 224.418 | |
| 2.819 | | | | 223.390 | |
| 1.028 | | | | 1.028 | checks |

* Permissible to start here because this is the only known RL; also, in working back to 1535 m one still subtracts from HPC in the usual way.

## Intermediate sight check

$$2245.723 = [(224.981 \times 7) + (226.393 \times 3) - (5.504 + 2.819)]$$
$$1574.867 + 679.179 - 8.323 = 2245.723$$

The student should now draw a sketch of the problem and add to it all the pertinent data as shown in *Figure 2.14*.

Examination of *Figure 2.14* shows the road to be rising from S to N at a regular grade of 0.510 m in 15 m. This infers then that the most northerly point (pt $B$ on east channel) should be the highest; however, as the crown of the road is higher than the channel one should also check pt $A$ on the crown; all other points can be ignored. Now, from the illustration the distance 1550 to $A$ on the centre-line

$$= 6 \times (2)^{1/2} = 8.5 \text{ m}$$

∴ Rise in level from 1550 to $A = (0.509/15) \times 8.5 = 0.288$ m
∴ Level at $A = 224.288$ m giving a clearance of $(229.547 - 224.288) = 5.259$ m
Distance 1550 to $B$ along east channel $= 8.5 + 4 = 12.5$ m
∴ Rise in level from 1550 to $B = (0.510/15) \times 12.5 = 0.425$ m
∴ Level at $B = 223.908 + 0.425 = 224.333$ m
∴ Clearance at $B = 229.547 - 224.333 = 5.214$ m
∴ Minimum clearance occurs at the most northerly point on the east channel, i.e. at $B$.

## Exercises

(*2.1*) The following readings were taken with a level and a 4.25-m staff:

0.683, 1.109, 1.838, 3.398 [3.877 and 0.451] CP, 1.405, 1.896, 2.676 BM
(102.120 AOD), 3.478 [4.039 and 1.835] CP, 0.649, 1.707, 3.722

Draw up a level book and reduce the levels by

(a) R-and-F.
(b) Height of collimation.

What error would occur in the final level if the staff had been wrongly extended and a
plain gap of 12 mm occurred at the 1.52-m section joint?                              (LU)

Parts (a) and (b) are self checking. Error in final level = zero.
(Hint: all readings greater than 1.52 m will be too small by 12 mm. Error in final level
will be calculated from BM only.)

(*2.2*) The following staff readings were observed (in the order given) when levelling up a
hillside from a TBM 135.2 m AOD. Excepting the staff position immediately after the
TBM, each staff position was higher than the preceding one.

1.408, 2.728, 1.856, 0.972, 3.789, 2.746, 1.597, 0.405, 3.280, 2.012, 0.625, 4.136, 2.664,
0.994, 3.901, 1.929, 3.478, 1.332

Enter the readings in level-book form by both the R-and-F and collimation systems
(these may be combined into a single form to save copying).                              (LU)

(*2.3*) The following staff readings in metres were obtained when levelling along the
centre-line of a straight road *ABC*.

| BS | IS | FS | Remarks |
|------|--------|-------|-----------------------------|
| 2.405 | | | pt *A* (RL = 250.05 m AOD) |
| 1.954 | | 1.128 | CP |
| 0.619 | | 1.466 | pt *B* |
| | 2.408 | | pt *D* |
| | −1.515 | | pt *E* |
| 1.460 | | 2.941 | CP |
| | | 2.368 | pt *C* |

*D* is the highest point on the road surface beneath a bridge crossing over the road at
this point and the staff was held inverted on the underside of the bridge girder at *E*,
immediately above *D*. Reduce the levels correctly by an approved method, applying the
checks, and determine the headroom at *D*. If the road is to be re-graded so that *AC* is a
uniform gradient, what will be the new headroom at *D*? The distance *AD* = 240 m and
*DC* = 60 m.                              (LU)

(*Answer:* 3.923 m, 5.071 m)

(*2.4*) Distinguish, in construction and method of use, between dumpy and tilting
levels. State in general terms the principle of an automatic level.                              (ICE)

(*2.5*) The following levels were taken with a metric staff on a series of pegs at 100-m intervals along the line of a proposed trench.

| BS | IS | FS | Remarks |
|------|------|------|-----------|
| 2.10 | | | TBM 28.75 m |
| | 2.85 | | peg *A* |
| 1.80 | | 3.51 | peg *B* |
| | 1.58 | | peg *C* |
| | 2.24 | | peg *D* |
| 1.68 | | 2.94 | peg *E* |
| | 2.27 | | |
| | 3.06 | | |
| | | 3.81 | TBM 24.07 m |

If the trench is to be excavated from peg *A* commencing at a formation level of 26.5 m and falling to peg *E* at a grade of 1 in 200; calculate the height of the sight rails in metres at *A*, *B*, *C*, *D* and *E*, if a 3-m boning rod is to be used.

(*Answer:* 1.50, 1.66, 0.94, 1.10, 1.30 m)

Briefly discuss the techniques and advantages of using laser beams for the control of more precise work.                                                     (KP)

Refer to Chapter 8 for details of 'sight rails' and 'lasers'.

## 2.7 PRECISE LEVELLING

The difference between precise and simple levelling lies in the use of more refined instruments and techniques for the former to attain much greater accuracy.

### 2.7.1 Definitions

In addition to the definitions for simple levelling the following are required:

(1) *Level line:* Imagine two points, *A* and *B*, some distance apart on the Earth's surface but having exactly the same level (*Figure 2.15*). Ignoring the effects of refraction and assuming the Earth to be a perfect sphere, the readings from *X'* on both staffs held at *A* and *B* would be identical. For this to be so, the line of sight would need to be curved parallel to the Earth's surface, giving readings at *A'* and *B'*. Such a line is called a *level line* and is at all points at right-angles to the direction of gravity.

(2) *Horizontal line:* In the above case, however, the line of sight would be from *X'* to *B''*, and this is called a *horizontal line*. The reading at *B''* would result in the level of *B* appearing too low by the amount *B'B''*. This error is due to the curvature of the Earth, and results in a *positive* correction of the amount *B'B''* to the apparent level of *B*.

The line *X'B''* does not, however, remain horizontal but is subject to refraction, and gives the actual staff reading at *Y*, the curvature correction is thus reduced by approximately one seventh.

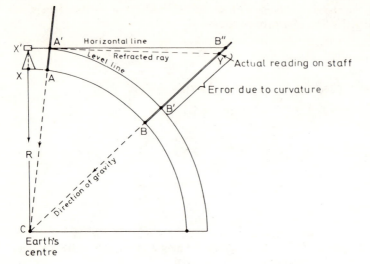

*Figure 2.15*

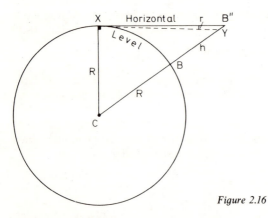

*Figure 2.16*

## 2.7.2 Curvature and refraction

### (1) Curvature

From *Figure 2.16*   $(XB'')^2 = (CB'')^2 - (CX)^2 = (R + h)^2 - R^2$
$$= R^2 + 2Rh + h^2 - R^2 = (2Rh + h^2)$$

Now, as the geodetic distance $XB$ is small compared with $R$ it can be taken as equal to $XB'' = D$.

$\therefore \quad D = (2Rh + h^2)^{1/2}$          (2.1)

or    $h = D^2/(2R + h)$                (2.2)

In practice, $h$ is small compared with $R$, therefore $h^2$ in equation (2.1) may be neglected, thus:

$D = (2Rh)^{1/2}$                   (2.3)

$h = D^2/2R$                    (2.4)

## (2) Refraction

This is a variable quantity changing with temperature, pressure, position on the Earth's surface, etc. It is generally taken as being one seventh of curvature and acting in the opposite direction.

Thus the combined correction for curvature and refraction

$$(BY) = (h - r) = (h - h/7) = 6h/7 = 6D^2/14R \qquad (2.5)$$
$$\text{and} \qquad D = (14Rh/6)^{1/2} = [7(2Rh)/6]^{1/2} \qquad (2.6)$$

Consider now the curvature correction for $D$ km taking $R = 6370$ km.

$$h = \frac{D^2}{2R} = \frac{(D \times 1000)^2}{2 \times 6370 \times 1000} = 0.0785D^2 \text{ m} \qquad (2.7)$$

where $D$ is in kilometres.

The combined correction for curvature and refraction is

$$\frac{6}{7} \times h = 0.0673D^2 \text{ m} \qquad (2.8)$$

Although our main concern here is for the value of $h$, questions frequently arise involving the calculation of distance $D$.

## Worked example

*Example 2.4.* In extending a triangulation survey of the mainland to a distant off-lying island, observations were made between two trig stations, the one 3000 m and the other 1000 m above sea level. If the ray from one station to the other grazed the sea, what was the approximate distance between stations, (a) neglecting refraction, and (b) allowing for it? ($R = 6400$ km). (ICE)

Refer *Figure 2.17*.

(a) $D_1 = (2Rh_1)^{1/2} = (2 \times 6400 \times 1)^{1/2} = 113$ km
$\quad D_2 = (2Rh_2)^{1/2} = (2 \times 6400 \times 3)^{1/2} = 196$ km
$\quad$ Total Distance $= 309$ km

(b) From equation (2.6): $D_1 = (7/6 \times 2R_1)^{1/2}$, $D_2 = (7/6 \times 2Rh_2)^{1/2}$.

By comparison with the equation in (a) above, it can be seen that the effect of refraction is to increase distance by $(7/6)^{1/2}$

$$\therefore \ D = 309 \times (7/6)^{1/2} = 334 \text{ km}$$

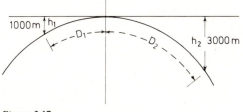

*Figure 2.17*

### 2.7.3 Reciprocal levelling

In precise levelling the lengths of the sights are kept equal to within 0.5 m. This has the effect of eliminating residual collimation error and errors due to curvature, and minimizing errors due to refraction.

When one needs to cross a wide gap in the process of levelling it becomes impossible to equalize the BS and FS lengths and so the method of *reciprocal levelling* is used.

With instrument near *A* (*Figure 2.18(a)*), the difference in level between

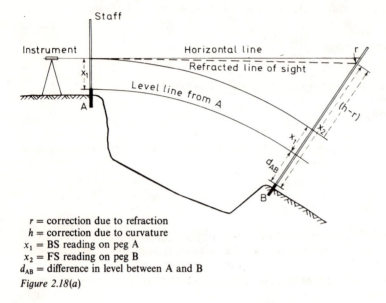

$r$ = correction due to refraction
$h$ = correction due to curvature
$x_1$ = BS reading on peg A
$x_2$ = FS reading on peg B
$d_{AB}$ = difference in level between A and B

*Figure 2.18(a)*

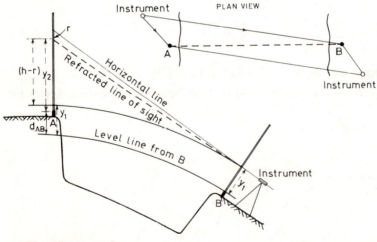

$y_1$ = BS reading on peg B
$y_2$ = FS reading on peg A

*Figure 2.18(b)*

$A$ and $B = d_{AB} = x_2 - x_1 - (h - r)$ $\qquad$ (2.9)

With instrument near $B$ (*Figure 2.18(b)*),

$d_{AB} = y_1 - [y_2 - (h - r)] = y_1 - y_2 + (h - r)$ $\qquad$ (2.10)

Now, $(x_2 - x_1)$ is the difference in the staff readings from $A = X$ and $(y_1 - y_2)$ is the difference in the staff readings from $B = Y$. Therefore, adding equations (2.9) and (2.10)

gives $\qquad 2d_{AB} = X + Y \qquad \therefore \; d_{AB} = \dfrac{X + Y}{2}$ $\qquad$ (2.11)

Thus, reciprocal levelling eliminates the effects of curvature and refraction, the difference in level between $A$ and $B$ simply being the mean of the difference in observations from each bank.

The above equation assumes the value of $r$ equal in both cases. However, if only one level is used there will be a time lag in transferring it to the opposite bank, during which the value of $r$ may change. Thus to ensure better results two levels, one on each bank should be used and the sights taken simultaneously. Although this will give better results than when using one level, each level may have a different collimation error. They should therefore be interchanged and the whole procedure repeated, the mean of the four values will then be the most probable difference in level between the two points.

## Worked examples

*Example 2.5.* Obtain, from first principles, an expression giving the combined correction for the Earth's curvature and atmospheric refraction in levelling, assuming that the Earth is a sphere of 12 740 km diameter. Reciprocal levelling between two points $Y$ and $Z$ 730 m apart on opposite sides of a river gave the following results:

| Instrument at | Height of instrument (m) | Staff at | Staff reading (m) |
|---|---|---|---|
| $Y$ | 1.463 | $Z$ | 1.688 |
| $Z$ | 1.436 | $Y$ | 0.991 |

Determine the difference in level between $Y$ and $Z$ and the amount of any collimation error in the instrument. $\qquad$ (ICE)

(1) $(h - r) = \dfrac{6D^2}{14R} = 0.0673D^2$ m $\qquad$ (Refer *Section 2.7.2*)

(2) With instrument at $Y$, $Z$ is lower by $(1.688 - 1.463) = 0.225$ m

With instrument at $Z$, $Z$ is lower by $(1.436 - 0.991) = 0.445$ m

*True* height of $Z$ below $Y = \dfrac{0.225 + 0.445}{2} = 0.335$ m

Instrument height at $Y = 1.463$ m; knowing now that $Z$ is lower by 0.335 m, then a truly horizontal reading on $Z$ should be $(1.463 + 0.335) = 1.798$ m; it was, however, 1.688 m, i.e. $-0.11$ m too low ($-$ve indicates low). This error is due to curvature and refraction $(h - r)$ *and* collimation error of the instrument $(e)$.

Thus:   $(h - r) + e = -0.110$ m

Now   $(h - r) = \dfrac{6D^2}{14R} = \dfrac{6 \times 730^2}{14 \times 6370 \times 1000} = 0.036$ m

$\therefore\ e = -0.110 - 0.036 = -0.146$ m in 730 m
$\therefore$ Collimation error $e = 0.020$ m *down* in 110 m.

*Example 2.6.* A and B are 2400 m apart. Observations with a level gave:

A, height of instrument 1.372 m, reading at B 3.359 m
B, height of instrument 1.402 m, reading at A 0.219 m

Calculate the difference of level and the error of the instrument, if refraction correction is one seventh that of curvature.                                          (LU)

Instrument at A, B is lower by $(3.359 - 1.372) = 1.987$ m
Instrument at B, B is lower by $(1.402 - 0.219) = \underline{1.183}$  m
                              *True* height of B below $A = 0.5 \times 3.170$ m $= 1.585$ m

Combined error due to curvature and refraction
   $= 0.0673D^2$ m $= 0.0673 \times 2.4^2 = 0.388$ m

Now using same procedure as in *Example 2.5.*

Instrument at $A = 1.372$, thus true reading at $B = (1.372 + 1.585)$
                                        $= 2.957$ m
            Actual reading at B               $= \underline{3.359}$  m
            Actual reading at B too high by    $+0.402$ m
Thus       $(h - r) + e = +0.402$ m
            $e = +0.402 - 0.388 = +0.014$ m in 2400 m
Collimation error   $e = +0.001$ m *up* in 100 m.

## Exercises

1(a) Determine from first principles the approximate distance at which correction for curvature and refraction in levelling amounts to 3 mm, assuming that the effect of refraction is one seventh that of the Earth's curvature and that the Earth is a sphere of 12 740 km diameter.
(b) Two survey stations A and B on opposite sides of a river are 780 m apart, and reciprocal levels have been taken between them with the following results:

| Instrument at | Height of instrument (m) | Staff at | Staff reading (m) |
|---|---|---|---|
| A | 1.472 | B | 1.835 |
| B | 1.496 | A | 1.213 |

Compute the ratio of refraction correction to curvature correction, and the difference in level between A and B: ((a) 210 m (b) 0.14 to 1; B lower by 0.323 m).

## 2.8 PRECISE LEVELLING EQUIPMENT

### 2.8.1 Staffs

*Precise levelling staffs* have a wooden frame carrying graduations on a strip of invar fixed at the bottom, but free to move along the remainder of the frame, thus allowing for thermal expansion. Some staffs have two invar strips, one graduated in reverse to eliminate gross reading errors. A built-in spirit bubble ensures verticality, and steadying rods are also supplied. The strip is divided into 10-mm or 5-mm intervals.

### Adjustments

(1)  The staff should be tested for verticality at least once a week, using a plumb bob and with the circular bubble adjusted if necessary.
(2)  Weekly tests should also be made for warping by stretching a fine wire from end to end. Maximum error should not be greater than 6 mm.
(3)  Graduation errors should be known and thus corrected by standardizing the staff against an invar tape. This is particularly important when two staves, each with a different graduation error, are used.

### 2.8.2 Levels

Precise levels are either tilting or automatic types, their accuracy depending mainly on the sensitivity of the bubble or the compensator and the magnification and resolution properties of the lenses.

(1) *Bubbles:* The greater the radius of curvature of the bubble tube the more sensitive is the bubble. Thus the bubble has a greater horizontal movement per degree of tilt. This makes any movement off centre more easily discernible.

   Bubbles are best horizontalized when viewed through a split-bubble system. This method is claimed to be eight times more accurate than when the bubble is viewed openly.

(2) *Magnification:* Directly increases the accuracy of readings on the staff. It also increases the sighting distance.

### 2.8.3 Parallel-plate micrometer

For precise levelling, the estimation of 1 mm is not sufficiently accurate. A parallel-plate glass micrometer in front of the object lens enables readings to be made direct to 0.1 mm, and estimated to 0.01 mm.

   The principle of the attachment is seen from *Figure 2.19*. Had the parallel plate been vertical the line of sight would have passed through without deviation and the reading would have been 1.026 m, the final figure being estimated. However, by manipulating the micrometer the parallel plate is tilted until the line of sight is displaced to the nearest indicated reading, which is 1.02 m. The amount of displacement *s* is measured on the micrometer and added to the exact reading to give 1.026 47 m, only the last decimal place being estimated.

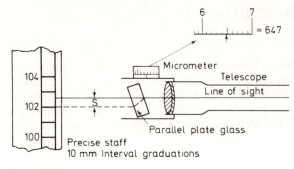

Figure 2.19

It can be seen from the figure that the plate could equally have moved in the opposite direction, displacing the line of sight up. To avoid the difficulty of whether to add or subtract *s*, the micrometer is always set to read zero before each sight. This will tilt the plate to its maximum position opposite to that shown in *Figure 2.19* and so displace the line of sight upwards. This will not affect the levelling provided that it is done for every sight. In this position the micrometer screw will move only from zero to ten, and the line of sight is always displaced down so *s* is always added.

Parallel-plate micrometers (PPM) are also manufactured for use with 5-mm graduations.

## 2.9 SOURCES OF ERROR

In addition to the sources of error already mentioned for both ordinary and precise work the following must be considered:

(1) To reduce errors due to staff and instrument sinking when levelling on soft ground, use levelling plates particularly on change points and obtain the sights quickly. To facilitate this, use two staffs and always sight to the same one first, as in *Figure 2.20*.

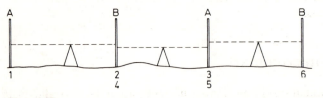

Figure 2.20

(2) Equalize lengths of sights to reduce the effects of curvature and refraction. In this case equalizing the sums of the lengths of BS and FS will not eliminate the error, as it is proportional to the distance squared.
(3) Lines of levels should be run forward in the morning and back in the evening, on the assumption that ground is colder in the morning and warmer in the evening, thus reducing refraction effects.

(4) Shield instrument from Sun's heat to reduce errors resulting from differential expansion of its parts.

(5) All circuits should be adjusted by the method of least squares. (Refer Volume 2, Chapter 1)

## 2.10 ACCURACY

As a guide to the acceptability or otherwise of the work, the variation in level obtained from two separate runs should not exceed $\pm 4(K)^{1/2}$ mm, where $K$ is the distance levelled in kilometres.

# 3

# Earthworks

Estimation of areas and volumes is basic to most engineering schemes such as route alignment, reservoirs, tunnels, etc. The excavation and hauling of material on such schemes is the most significant and costly aspect of the work, on which profit or loss may depend.

Areas may be required in connection with the purchase or sale of land, with the sub-division of land or with the grading of land.

Earthwork volumes must be estimated to enable route alignment to be located at such lines and levels that cut and fill are balanced as far as practicable; and to enable contract estimates of time and cost to be made for proposed work; and to form the basis of payment for work carried out.

The tedium of earthwork computation has now been removed by the use of micro and main-frame computers. Digital ground models (DGM), in which the ground surface is defined mathematically in terms of $x$, $y$ and $z$ co-ordinates, are now stored in the computer memory. This data bank may now be used with several alternative design schemes to produce the optimum route in both the horizontal and vertical planes. In addition to all the setting-out data, cross-sections are produced, earthwork volumes supplied and mass-haul diagrams drawn. Quantities may be readily produced for tender calculations and project planning. The data banks may be updated with new survey information at any time and further facilitate the planning and management not only of the existing project but of future ones.

However, before the impact of modern computer technology can be realized, one requires a knowledge of the fundamentals of areas and volumes, not only to produce the software necessary, but to understand the input data required and to be able to interpret and utilize the resultant output properly.

## 3.1 AREAS

The computation of areas may be based on data scaled from plans or drawings; or direct from the survey field data.

### 3.1.1 Plotted areas

(1) It may be possible to sub-divide the plotted area into a series of triangles, measure the sides $a, b, c$, and compute the areas using:

$$\text{Area} = [s(s-a)(s-b)(s-c)]^{1/2} \qquad \text{where } s = (a+b+c)/2$$

The accuracy achieved will be dependent upon the scale error of the plan and the accuracy to which the sides are measured.

(2) Where the area is irregular, a sheet of gridded tracing material may be superimposed over it and the number of squares counted. Knowing the scale of the plan and the size of the squares, an estimate of the area can be obtained. Portions of squares cut by the irregular boundaries can be estimated.

(3) Alternatively, irregular boundaries may be reduced to straight lines using *give-and-take lines*, in which the areas 'taken' from the total area balance out with the extra areas 'given' (*Figure 3.1*).

Take

Give

A r e a

*Figure 3.1*

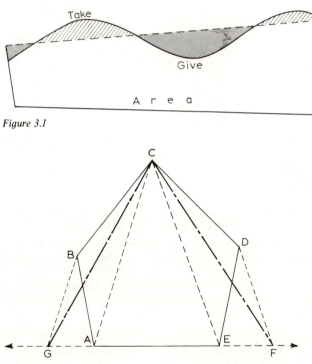

*Figure 3.2*

(4) If the area is a polygon with straight sides it may be reduced to a triangle of equal area. Consider the polygon *ABCDE* shown in *Figure 3.2*.

Take *AE* as the base and extend it as shown. Join *CE* and from *D* draw a line parallel to *CE* on to the base at *F*. Similarly, join *CA* and draw a line parallel from *B* on to the base at *G*. Triangle *GCF* has the same area as the polygon *ABCDE*.

(5) The most common method of measuring areas from plans is to use an instrument called a *planimeter* (*Figure 3.3(a)*). This comprises two arms, *JF* and *JP*, which are free to move relative to each other through the hinged joint at *J* but fixed to the plan by a weighted needle at *F*. *M* is the graduated measuring wheel and *P* the tracing point. As *P* is moved around the perimeter of the area the measuring wheel partly rotates, partly slides over the plan with the varying movement of the tracing point (*Figure 3.3(b)*). The

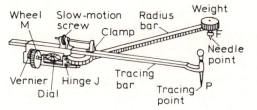

*Figure 3.3(a)* Amsler's polar planimeter

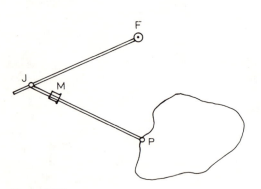

*Figure 3.3(b)*

measuring wheel is graduated circumferentially into 10 divisions, each of which is further sub-divided by 10 into one-hundredths of a revolution, whilst a vernier enables readings to one thousandth of a revolution. The wheel is connected to a dial which records the numbered revolutions up to 10. On a *fixed-arm planimeter* one revolution of the wheel may represent 100 mm² on a 1:1 basis; thus, knowing the number of revolutions and the scale of the plan, the area is easily computed. In the case of a *sliding-arm planimeter* the sliding arm $JP$ may be set to the scale of the plan, thereby facilitating more direct measurement of the area.

In the normal way, needle point $F$ is fixed *outside* the area to be measured, the initial reading noted, the tracing point traversed around the area and the final reading noted. The difference of the two readings gives the number of revolutions of the measuring wheel, which is a direct measure of the area. If the area is too large to enable the whole of its boundary to be traversed by the tracing point $P$ when the needle point $F$ is outside the area, then the area may be sub-divided into smaller more manageable areas, or the needle point can be transposed *inside* the area.

As the latter procedure requires the application of the *zero circle* of the instrument, the former approach is preferred.

The zero circle of a planimeter is that circle described by the tracing point $P$, when the needle point $F$ is at the centre of the circle, and the two arms $JF$ and $JP$ are at right-angles to each other. In this situation the measuring wheel is normal to its path of movement and so slides without rotation, thus producing a zero change in reading. The value of the zero circle is supplied with the instrument.

If the area to be measured is greater than the zero circle (*Figure 3.4(a)*) then only the

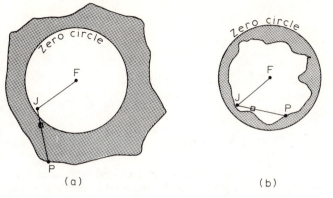

(a)                                         (b)

*Figure 3.4*

tinted area is measured, and the zero circle value must be added to the difference between the initial and final wheel readings. In such a case the final reading will always be *greater* than the initial reading. If the final reading is *smaller* than the initial reading, then the situation is as shown in *Figure 3.4(b)* and the measured area, shown tinted, must be subtracted from the zero circle value.

## Example 3.1

(a) *Forward movement*                      (b) *Backward movement*

| | | | | |
|---|---|---|---|---|
| Initial reading | 2.497 | | Initial reading | 2.886 |
| Final reading | 6.282 | | Final reading | 1.224 |

| | | | | |
|---|---|---|---|---|
| Difference | 3.785 revs | | Difference | 1.662 revs |
| Add zero circle | 18.546 | | Subtract from | |
| | | | zero circle | 18.546 |
| Area = 22.231 revs | | | | |
| | | | Area = 16.884 revs | |

If one revolution corresponds to an area of $(A)$, then on a plan of scale 1 in $M$, the actual area in (a) above equals $22.331 \times A \times M^2$.

(6) If the shape of the area is defined in terms of plane rectangular co-ordinates, its area can be computed precisely, as shown in *Section 4.7* (Chapter 4).

(7) If the area can be divided into strips then the area can be found using either (a) the trapezoidal rule or (b) Simpson's rule, as follows (*Figure 3.5*).

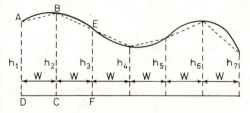

*Figure 3.5* Trapezoidal and Simpson's rules

(a) *Trapezoidal rule*

Area of 1st trapezoid $ABCD = \dfrac{h_1 + h_2}{2} \times w$

Area of 2nd trapezoid $BEFC = \dfrac{h_2 + h_3}{2} \times w$ and so on.

$\therefore$ Total area = sum of trapezoids

$$= A = w\left(\frac{h_1 + h_7}{2} + h_2 + h_3 + h_4 + h_5 + h_6\right) \tag{3.1}$$

*N.B.* (i) If the first or last ordinate is zero, it must still be included in the equation.
(ii) The formula represents the area bounded by the broken line under the curving boundary; thus, if the boundary curves outside then the computed area is too small, and *vice versa*.

(b) *Simpson's rule*

$$A = w[(h_1 + h_7) + 4(h_2 + h_4 + h_6) + 2(h_3 + h_5)]/3 \tag{3.2}$$

i.e. one third the distance between ordinates, multiplied by the sum of the *first* and *last* ordinates, plus four times the sum of the *even* ordinates, plus twice the sum of the *odd* ordinates.

*N.B.* (i) This rule assumes a curved boundary and is therefore more accurate than the trapezoidal rule. If the boundary was a parabola the formula would be exact.
(ii) The equation requires an *odd* number of ordinates and consequently an even number of areas.

The above equations are also useful for calculating areas from chain survey data. The areas enclosed by the chain lines are usually in the form of triangles, whilst the offsets to the irregular boundaries become the ordinates for use with the equations.

### 3.1.2 Cross-sections

Finding the areas of cross-sections is the first step in obtaining the volume of earthwork to be handled in route alignment projects (road or railway), or reservoir construction, for examples.

In order to illustrate more clearly what is meant by the above statement, let us consider a road construction project. In the first instance an accurate plan is produced on which to design the proposed route. The centre-line of the route, defined in terms of rectangular co-ordinates at 10- to 30-m intervals, is then set out in the field. Ground levels are obtained along the centre-line and also at right-angles to the line (*Figure 3.6(a)*). The levels at right-angles to the centre-line depict the ground profile, as shown in *Figure 3.6(b)*, and if the design template, depicting the formation level, road width, camber, side slopes, etc. is added, then a cross-section is produced whose area can be obtained by planimeter or computation. The shape of the cross-section is defined in terms of vertical heights (levels) at horizontal distances each side of the centre-line, thus no matter how complex the shape, these parameters can be treated as rectangular co-ordinates and the area computed using the rules given in Chapter 4. The areas may now

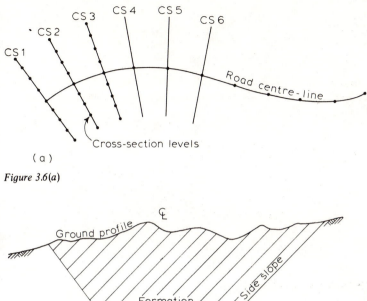

(a)

*Figure 3.6(a)*

(b)

*Figure 3.6(b)* Cross-sectional area of cutting

be used in various rules (see later) to produce an estimate of the volumes. Levels along, and normal to, the centre-line may be obtained by standard levelling procedures, by optical or electromagnetic tacheometry, or by aerial photogrammetry. The whole computational procedure, including the road design and optimization, would then be carried out on the computer to produce volumes of cut and fill, accumulated volumes, areas and volumes of top-soil strip, side widths, etc. Where plotting facilities are available the program would no doubt include routines to plot the cross-sections for visual inspection.

An example of an earthworks program, written in standard basic, is given in Appendix A on page 286.

Where there are no computer facilities the cross-sections may be approximated to the ground profile to afford easy computation. The particular cross-section adopted would be dependent upon the general shape of the ground. Typical examples are illustrated in *Figure 3.7.*

Whilst equations are available for computing the areas and side widths they tend to be over-complicated and the following method using 'rate of approach' is recommended (*Figure 3.8*).

*Given:* height $x$ and grades $AB$ and $CB$ in triangle $ABC$.
*Required:* to find distance $y_1$.
*Method:* Add the two grades, using their absolute values, invert them and multiply by $x$.

i.e. $(1/5 + 1/2)^{-1}x = 10x/7 = y_1$

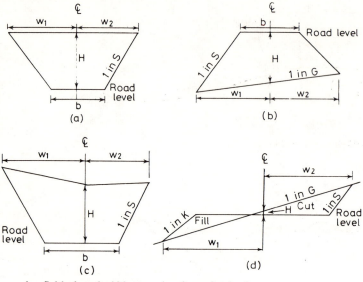

$b$ = finished road width at road or formation level
$H$ = centre height
$W_1, W_2$ = side widths, measured horizontally from the centre-line and depicting the limits of the construction
1 in $S$ = side slope of 1 vertical to $S$ horizontal
1 in $G$ = existing ground slope

*Figure 3.7* (*a*) Cutting, (*b*) embankment, (*c*) cutting and (*d*) hillside section

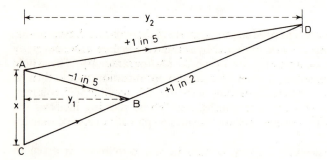

*Figure 3.8* Rate of approach

Similarly, to find distance $y_2$ in triangle $ADS$. *Subtract* the two grades, invert them and multiply by $x$.

e.g.   $(1/5 - 1/2)^{-1}x = 10x/3 = y_2$

The rule, therefore, is:

(1) When the two grades are running in opposing directions (as in $ABC$), *add* (signs opposite $+ -$).
(2) When the two grades are running in the same direction (as in $ABD$), *subtract* (signs same).

*N.B.* Height $x$ must be *vertical* relative to the grades (see *Worked example 3.8*, p. 65).

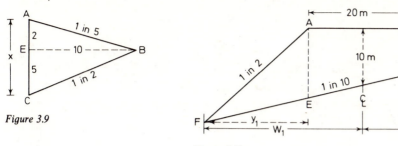

Figure 3.9

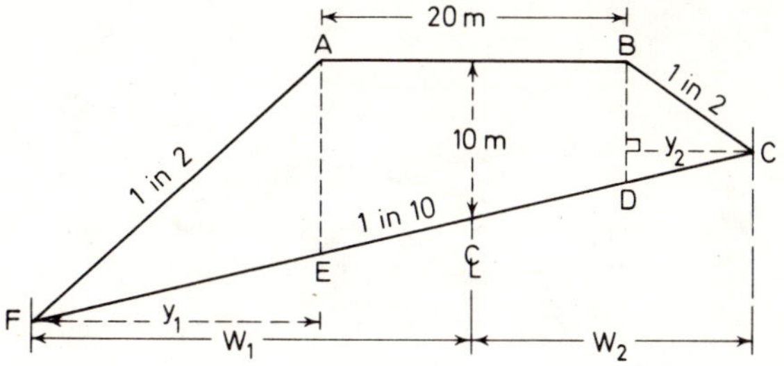

Figure 3.10

## Proof

From *Figure 3.9* it is seen that 1 in 5 = 2 in 10 and 1 in 2 = 5 in 10, thus the two grades diverge from $B$ at the rate of 7 in 10. Thus, if $AC = 7$ m then $EB = 10$ m, i.e. $x \times 10/7 = 7 \times 10/7 = 10$ m.

Two examples will now be worked to illustrate the use of the above technique.

## Worked examples

*Example 3.2.* Calculate the *side widths* and *cross-sectional area* of an embankment (*Figure 3.10*) having the following dimensions:

Road width = 20 m     existing ground slope = 1 in 10
Side slopes = 1 in 2     centre height = 10 m

As horizontal distance from centre-line to $AE$ is 10 m and the ground slope is 1 in 10, then $AE$ will be 1 m greater than the centre height and $BD$ 1 m less. Thus, $AE = 11$ m and $BD = 9$ m, area of $ABDE = 20 \times 10 = 200$ m$^2$. Now, to find the areas of the remaining triangles $AEF$ and $BDC$ one needs the perpendicular heights $y_1$ and $y_2$, as follows:

(1)  $1/2 - 1/10 = 4/10$, then $y_1 = (4/10)^{-1} \times AE = 11 \times 10/4 = 27.5$ m
(2)  $1/2 + 1/10 = 6/10$, then $y_2 = (6/10)^{-1} \times BD = 9 \times 10/6 = 15.0$ m

$\therefore$ Area triangle $AEF = \dfrac{AE}{2} \times y_1 = \dfrac{11}{2} \times 27.5 = 151.25$ m$^2$

Area triangle $BDC = \dfrac{BD}{2} \times y_2 = \dfrac{9}{2} \times 15.0 = 67.50$ m$^2$

Total area = $(200 + 151.25 + 67.5) = 418.75$ m$^2$
Side width $w_1 = 10$ m $+ y_1 = 37.5$ m
Side width $w_2 = 10$ m $+ y_2 = 25.0$ m

*Example 3.3.* Calculate the side widths and cross-sectional areas of cut and fill on a hillside section (*Figure 3.11*) having the following dimensions:

Road width = 20 m     existing ground slope = 1 in 5
Side slope in cut = 1 in 1     centre height in cut = 1 m
Side slope in fill = 1 in 2.

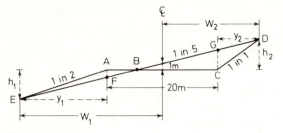

*Figure 3.11*

As ground slope is 1 in 5 and centre height 1 m, it follows that the horizontal distance from centre-line to $B$ is 5 m; therefore, $AB = 5$ m, $BC = 15$ m. From these latter distances it is obvious that $AF = 1$ m and $GC = 3$ m.

Now,   $y_1 = (1/2 - 1/5)^{-1} \times AF = \dfrac{10}{3} \times 1 = 3.3$ m

$$y_2 = (1 - 1/5)^{-1} \times GC = \frac{5}{4} \times 3 = 3.75 \text{ m}$$

∴ Side width $w_1 = 10$ m $+ y_1 = 13.3$ m
  Side width $w_2 = 10$ m $+ y_2 = 13.75$ m

Now, as side slope $AE$ is 1 in 2, then $h_1 = y_1/2 = 1.65$ m and as side slope $CD$ is 1 in 1, then $h_2 = y_2 = 3.75$ m

∴  Area of cut $(BCD) = \dfrac{BC}{2} \times h_2 = \dfrac{15}{2} \times 3.75 = 28.1 \text{ m}^2$

   Area of fill $(ABE) = \dfrac{AB}{2} \times h_1 = \dfrac{5}{2} \times 1.65 = 4.1 \text{ m}^2$

The student is now advised to compute the area and side widths of *Figure 3.12*, using the above techniques.

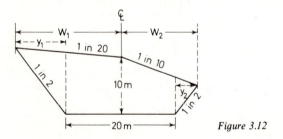

*Figure 3.12*

($y_1 = 23.3$, $w_1 = 33.3$, $y_2 = 15$, $w_2 = 25$, area $= 387.3$)

## 3.1.3 Dip and strike

On a tilted plane there is a direction of maximum tilt, such direction being called *the line of full dip*. Any line at right-angles to full dip will be a level line and is called a *strike line*

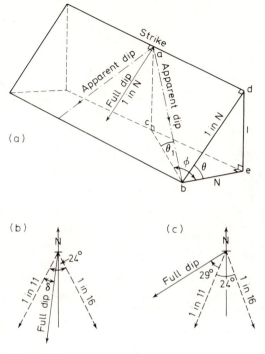

Figure 3.13

(*Figure 3.13(a)*). Any grade between full dip and strike is called *apparent dip*. An understanding of dip and strike is occasionally necessary for some earthwork problems. From *Figure 3.13(a)*

$$\tan \theta_1 = \frac{ac}{bc} = \frac{de}{bc} = \left(\frac{de}{be} \times \frac{be}{bc}\right) = \tan \theta \cos \phi$$

i.e.   tan (apparent dip) = tan (full dip) × cos (included angle)          (3.3)

## Worked example

*Example 3.4.* On a stratum plane, an apparent dip of 1 in 16 bears 170°, whilst the apparent dip in the direction 194° is 1 in 11; calculate the direction and rate of full dip.

Draw a sketch of the situation (*Figure 3.13(b)*) and assume any position for full dip. Now, using equation (3.3)

$$\tan \theta_1 = \tan \theta \cos \phi$$

$$\frac{1}{16} = \tan \theta \cos (24° - \delta)$$

$$\tan \theta = \frac{1}{16 \cos (24° - \delta)} \qquad\qquad (a)$$

Similarly,    $\dfrac{1}{11} = \tan \theta \cos \delta$

$$\tan \theta = \frac{1}{11 \cos \delta} \tag{b}$$

Equating (a) and (b)

$$16 \cos (24° - \delta) = 11 \cos \delta$$

$$16(\cos 24° \cos \delta + \sin 24° \sin \delta) = 11 \cos \delta$$

$$16(0.912 \cos \delta + 0.406 \sin \delta) = 11 \cos \delta$$

$$14.6 \cos \delta + 6.5 \sin \delta = 11 \cos \delta$$

$$3.6 \cos \delta = -6.5 \sin \delta$$

Cross multiply    $\dfrac{\sin \delta}{\cos \delta} = \tan \delta = -\dfrac{3.6}{6.5}$

$$\therefore \ \delta = -29°$$

*N.B.* The minus sign indicates that the initial position for full dip in *Figure 3.13(b)* is incorrect, and that it lies *outside* the apparent dip. As the grade is increasing from 1 in 16 to 1 in 11, the full dip must be as in *Figure 3.13(c)*.

$$\therefore \ \text{Direction of full dip} = 223°$$

Now, a second application of the formula will give the rate of full dip. That is

$$\frac{1}{11} = \frac{1}{x} \cos 29°$$

$$\therefore \ x = 11 \cos 29° = 9.6$$

$$\therefore \ \text{Rate of full dip} = 1 \text{ in } 9.6$$

## 3.2 VOLUMES

The importance of volume assessment has already been outlined. Many volumes encountered in civil engineering appear, at first glance, to be rather complex in shape. Generally speaking, however, they can be divided into *prisms*, *wedges* or *pyramids*, each of which will now be dealt with in turn.

### (1) Prism

The two ends of the prism (*Figure 3.14*) are equal and parallel, the resulting sides thus being parallelograms.

$$\text{Vol} = AL \tag{3.4}$$

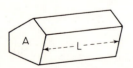

*Figure 3.14* Prism

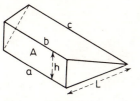

Figure 3.15 Wedge

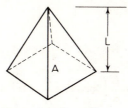

Figure 3.16 Pyramid

## (2) Wedge

Volume of wedge (*Figure 3.15*) $= \dfrac{L}{6}$ (sum of parallel edges × vertical height of base)

$$= \frac{L}{6}[(a + b + c) \times h] \tag{3.5a}$$

when $a = b = c$:          $V = AL/2.$ $\tag{3.5b}$

## (3) Pyramid

Volume of pyramid (*Figure 3.16*) $= \dfrac{AL}{3}$ $\tag{3.6}$

Equations (3.4) to (3.6) can all be expressed as the common equation:

$$V = \frac{L}{6}(A_1 + 4A_m + A_2) \tag{3.7}$$

where $A_1$ and $A_2$ are the end areas and $A_m$ is the area of the section situated midway between the end areas. It is important to note that $A_m$ is not the arithmetic mean of the end areas, except in the case of a wedge.

To prove the above statement consider

## (1) Prism

In this case $A_1 = A_m = A_2$ (*Figure 2.10*)

$$V = \frac{L}{6}(A + 4A + A) = \frac{L \times 6A}{6} = AL$$

## (2) Wedge

In this case $A_m$ is the mean of $A_1$ and $A_2$, but $A_2 = 0$. Thus $A_m = A/2$

$$V = \frac{L}{6}\left(A + 4 \times \frac{A}{2} + 0\right) = \frac{L \times 3A}{6} = \frac{AL}{2}$$

## (3) Pyramid

In this case $A_m = \dfrac{A}{4}$ and $A_2 = 0$

$$V = \frac{L}{6}\left(A + 4 \times \frac{A}{4} + 0\right) = \frac{L \times 2A}{6} = \frac{AL}{3}$$

Thus, any solid which is a combination of the above three forms and having a common value for $L$, may be solved using equation (3.7). Such a volume is called a *prismoid* and the formula is called the *prismoidal equation*. It is easily deduced by simply substituting areas for ordinates in Simpson's rule. The prismoid differs from the prism in that its parallel ends are not necessarily equal in area; the sides are generated by straight lines from the edges of the end areas (*Figure 3.17*).

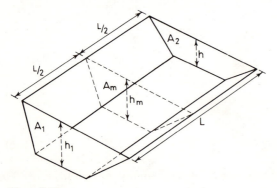

*Figure 3.17*

The prismoidal equation is correct when the figure is a true prismoid. In practice it is applied by taking three successive cross-sections. If the mid-section is different from that of a true prismoid, then errors will arise. Thus in practice sections should be chosen in order to avoid this fault. Generally, the engineer elects to observe cross-sections at regular intervals assuming compensating errors over a long route distance.

### 3.2.1 End-area method

Consider *Figure 3.18*, then

$$V = \frac{A_1 + A_2}{2} \times L \tag{3.8}$$

i.e. the mean of the two end areas multiplied by the length between them. This equation is correct only when the mid-area of the prismoid is the mean of the two end areas. It is correct for wedges but in the case of a pyramid it gives a result which is 50% too great

$$\text{Vol of pyramid} = \frac{A + 0}{2} \times L = \frac{AL}{2} \qquad \text{instead of} \qquad \frac{AL}{3}$$

Although this method generally over-estimates, it is widely used in practice. The

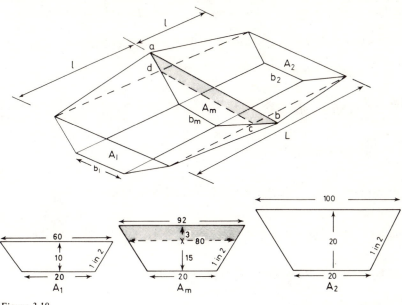

Figure 3.18

main reasons for this are its simplicity and the fact that the assumptions required for a good result using the prismoidal method are rarely fulfilled in practice. Strictly, however, it should be applied to prismoids comprising prisms and wedges only; such is the case where the height or width of the consecutive sections is approximately equal. It is interesting to note that with consecutive sections, where the height increases as the width decreases, or *vice versa*, the end-area method gives too small a value.

The difference between the prismoidal and end-area equations is called *prismoidal excess* and may be applied as a correction to the end-area value. It is rarely used in practice.

Summing a series of end areas gives:

$$V = L\left(\frac{A_1 + A_n}{2} + A_2 + A_3 + \cdots + A_{n-1}\right) \tag{3.9}$$

called the *trapezoidal rule* for volumes.

## 3.2.2 Comparison of end-area and prismoidal equations

In order to compare the methods, the volume of *Figure 3.18* will be computed as follows:

Dimensions of *Figure 3.18*.

Centre heights: $h_1 = 10$ m, $h_2 = 20$ m, $h_m = 18$ m
Road widths: $b_1 = b_2 = b_m = 20$ m
Side slopes: 1 in 2
Horizontal distance between sections: $l = 30$ m, $L = 60$ m

N.B. For a true prismoid $h_m$ would have been the mean of $h_1$ and $h_2$, equal to 15 m. The broken line indicates the true prismoid, the excess area of the mid-section is shown tinted.

The true volume is thus a true prismoid plus two wedges, as follows:

(1) $A_1 = \dfrac{60 + 20}{2} \times 10 = 400 \text{ m}^2$

$A_2 = \dfrac{100 + 20}{2} \times 20 = 1200 \text{ m}^2$

$A_m = \dfrac{80 + 20}{2} \times 15 = 750 \text{ m}^2$

Vol of prismoid $= V_1 = \dfrac{60}{6}(400 + 4 \times 750 + 1200) = 46\,000 \text{ m}^3$

Vol of wedge 1 $= \dfrac{L}{6}[(a + b + c) \times h] = \dfrac{30}{6}[(92 + 80 + 60) \times 3] = 3480 \text{ m}^3$

Vol of wedge 2 $= \dfrac{30}{6}[(92 + 80 + 100) \times 3] = 4080 \text{ m}^3$

*Total true volume = 53 560 m³*

(2) *Volume by prismoidal equation* ($A_m$ will now have a centre height of 18 m)

$A_m = \dfrac{92 + 20}{2} \times 18 = 1008 \text{ m}^2$

$\text{Vol} = \dfrac{60}{6}(400 + 4032 + 1200) = 56\,320 \text{ m}^3$

Error $= 56\,320 - 53\,560 = +2760 \text{ m}^2$

This error is approximately equal to the area of the excess mid-section multiplied by $\dfrac{L}{6}$, i.e. $\dfrac{\text{Area } abcd \times L}{6}$, and is so for all such circumstances; it would be $-$ve if the mid-area had been smaller.

(3) *Volume by end area*

$V_1 = \dfrac{400 + 1008}{2} \times 30 = 21\,120 \text{ m}^3$

$V_2 = \dfrac{1008 + 1200}{2} \times 30 = 33\,120 \text{ m}^3$

Total vol $= 54\,240 \text{ m}^3$

Error $= 54\,240 - 53\,560 = +680 \text{ m}^3$

Thus, in this case the end-area method gives a better result than the prismoidal equation. However, if we consider only the true prismoid, the volume by end areas is 46 500 m³ compared with the volume by prismoidal equation of 46 000 m³, which, in this case, is the true volume.

Therefore, in practice, it can be seen that neither of these two methods is satisfactory. Unless the ideal geometric conditions exist, which is rare, both methods will give errors. To achieve greater accuracy, the cross-sections should be located in the field, with due regard to the formula to be used. If the cross-sections are approximately equal in size and shape, and the intervening surface roughly a plane, then end areas will give an acceptable result. Should the sections be vastly different in size and shape, with the mid-section contained approximately by straight lines generated between the end sections, then the prismoidal equation will give the better result.

### 3.2.3 Contours

Volumes may be found from contours using either the end-area or prismoidal method. The areas of the sections are the areas encompassed by the contours. The distance between the sections is the contour interval. This method is commonly used for finding the volume of a reservoir, lake or spoil heap (refer to *Exercise (3.3)*, p. 67).

### 3.2.4 Spot heights

This method is generally used for calculating the volumes of excavations for basements or tanks, i.e. any volume where the sides and base are planes, whilst the surface is broken naturally (*Figure 3.19(a)*). *Figure 3.19(b)* shows the limits of the excavation with surface levels in metres at $A$, $B$, $C$ and $D$. The sides are vertical to a formation level of 20 m. If the area $ABCD$ was a plane then the volume of excavation would be:

$$V = \text{plan area } ABCD \times \text{mean height} \qquad (3.11)$$

However, as the illustration shows, the surface is very broken and so must be covered with a grid such that the area within each 10-m grid square is approximately a plane. It is therefore the ruggedness of the ground that controls the grid size. If, for instance, the surface $Aaed$ was not a plane, it could be split into two triangles by a diagonal ($Ae$) if this would produce better surface planes.

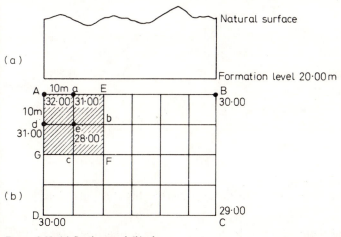

*Figure 3.19* (a) Section, and (b) plan

Considering square *Aaed* only:

$V$ = plan area × mean height

$$= 100 \times \frac{1}{4}(12 + 11 + 8 + 11) = 1050 \text{ m}^3$$

If the grid squares are all equal in area, then the data is easily tabulated and worked as follows:

Considering *AEFG* only, instead of taking each grid square separately, one can treat it as a whole.

$$\therefore \ V = \frac{100}{4}\left[h_A + h_E + h_F + h_G + 2(h_a + h_b + h_c + h_d) + 4h_e\right]$$

If one took each grid separately it would be seen that the heights of *AEFG* occur only once, whilst the heights of *abcd* occur twice and $h_e$ occurs four times; one still divides by four to get the mean height.

The above formula is very useful for any difficult shape consisting entirely of planes, as the following example illustrates (*Figures 3.20* and *3.21*).

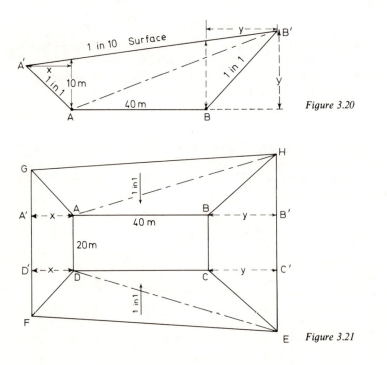

Figure 3.20

Figure 3.21

Vertical height at *A* and *D* is 10 m.

As *AB* = 40 m and surface slopes at 1 in 10, then vertical heights at *B* and *C* must be 4 m greater, i.e. 14 m.

Consider splitting the shape into two wedges by a plane connecting *AD* to *HE*.

In $\triangle ABB'$ (*Figure 3.20*):

By rate of approach: $y = \left(1 - \dfrac{1}{10}\right)^{-1} \times 14 = 15.56 \text{ m} = B'H = C'E$

$$\therefore\ HE = 20 + 15.56 + 15.56 = 51.12 \text{ m}.$$

Area of $\triangle ABB'$ normal to $AD, BC, HE$

$$= \frac{40}{2} \times 15.56 = 311.20 \text{ m}^2$$

$\therefore$ Vol = area × mean height $= \dfrac{311.20}{3}(AD + BC + HE)$

$$= 103.73(20 + 20 + 51.12) = 9452 \ m^3$$

Similarly in $\triangle AA'B'$:

$$x = \left(1 + \frac{1}{10}\right)^{-1} \times 10 = 9.09 \text{ m} = A'G = D'F.$$

$$\therefore\ GF = 20 + 9.09 + 9.09 = 38.18 \text{ m}.$$

Area of $\triangle AA'B'$ normal to $AD, GF, HE$

$$= \frac{(x + AB + y)}{2} \times 10 = \frac{64.65}{2} \times 10 = 323.25 \text{ m}^2$$

$$\therefore\ \text{Vol} = \frac{323.25}{3}(20 + 38.18 + 51.12) = 11\,777 \text{ m}^3$$

*Total vol = 21 229 $m^3$*

*Check*

Wedge $ABB' = \dfrac{40}{6}[(20 + 20 + 51.12) \times 15.56] = 9452 \text{ m}^3$

Wedge $AA'B' = \dfrac{64.65}{6}[(20 + 38.18 + 51.12) \times 10] = 11\,777 \text{ m}^3$

## 3.2.5 Effect of curvature on volumes

The application of the prismoidal and end-area formulae has assumed, up to this point, that the cross-sections are parallel. When the excavation is curved (*Figure 3.22*), the sections are radial and a *curvature correction* must be applied to the formulae.

Pappus's theorem states that the correct volume is where the distance between the cross-sections is taken along the path of the centroid.

Consider the volume between the first two sections of area $A_1$ and $A_2$:

Distance between sections measured along centre-line = $X'Y' = D$.
Angle $\delta$ subtended at the centre = $D/R$ radians.

Now, length along path of centroid = $XY = \delta \times$ mean radius to path of centroid where mean radius

$$= R - (d_1 + d_2)/2 = (R - d)$$

$$\therefore\ XY = \delta(R - d) = D(R - d)/R.$$

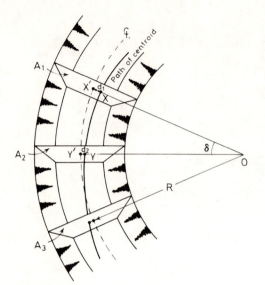

Figure 3.22

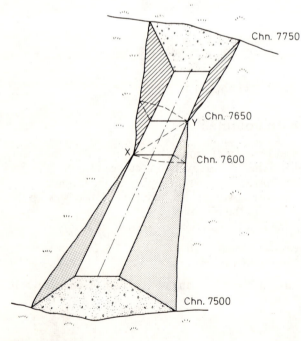

Figure 3.23

Vol by end areas $= \frac{1}{2}(A_1 + A_2)XY = \frac{1}{2}(A_1 + A_2)D(R - d)/R.$

$$= \frac{1}{2}(A_1 + A_2)D(1 - d/R)$$

In other words, one corrects for curvature by multiplying the area $A_1$ by $(1 - d_1/R)$, and area $A_2$ by $(1 - d_2/R)$, the corrected areas then being used in either the end-area or prismoidal formulae, in the normal way, with $D$ being the distance measured along the centre-line. If the centroid lay beyond the center-line, as in section $A_3$, then the correction is $(1 + d_3/R)$.

This correction for curvature, is again, never applied to earthworks in practice. Indeed it can be shown that the effect is cancelled out on long earthwork projects. However, it may be significant on small projects or single curved excavations.

## Worked examples

*Example 3.5. Figure 3.23* illustrates a section of road construction to a level road width of 20 m, which includes a change from fill to cut. From the data supplied in the following field book extract, calculate the volumes of cut and fill using the end-area method and correcting for prismoidal excess. (KP)

| Chainage | Left | Centre | Right |
|---|---|---|---|
| 7500 | $-\frac{10.0}{36.0}$ | $-\frac{20.0}{0}$ | $-\frac{8.8}{22.0}$ |
| 7600 | $\frac{0}{10}$ | $-\frac{6.0}{0}$ | $-\frac{14.0}{24.6}$ |
| 7650 | $\frac{16.0}{22.0}$ | $\frac{4.0}{0}$ | $\frac{0}{10}$ |
| 7750 | $\frac{13.5}{24.0}$ | $\frac{22.0}{0}$ | $\frac{8.6}{26.0}$ |

N.B. (1) Students should note the method of booking and compare it with the cross-sections in *Figure 3.24*.

(2) The method of splitting the sections into triangles for easy computation should also be noted.

Area of cross-section 75 + 00

area $\Delta 1 = \dfrac{10 \times 10}{2} = 50 \ m^2$

area $\Delta 2 = \dfrac{36 \times 20}{2} = 360 \ m^2$

area $\Delta 3 = \dfrac{22 \times 20}{2} = 220 \ m^2$

area $\Delta 4 = \dfrac{8.8 \times 10}{2} = 44 \ m^2$

*Total area* $= 674 \ m^2$

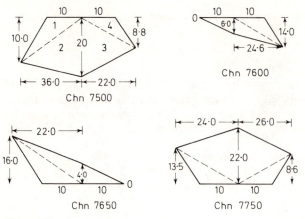

Figure 3.24

Similarly, area of cross-section $76 + 00 = 173.8$ m$^2$

$$\text{Vol by end area} = \frac{674 + 173.8}{2} \times 100 = 42\,390 \text{ m}^3$$

The equation for prismoidal excess varies with the shape of the cross-section. In this particular instance it equals

$$\frac{L}{12}(H_1 - H_2)(W_1 - W_2)$$

where    $L$ = horizontal distance between the two end areas
           $H$ = centre height
           $W$ = the sum of the side widths per section, i.e. $(w_1 + w_2)$

Thus, prismoidal excess $= \dfrac{100}{12}(20 - 6)(58 - 34.6) = \quad 2\,730$ m$^3$

$$\text{\underline{\hspace{3cm}}}$$

Corrected volume $= 39\,660$ m$^3$

Vol between $76 + 00$ and $76 + 50$

Line $XY$ in *Figure 3.23* shows clearly that the volume of fill in this section forms a pyramid with the cross-section $76 + 00$ as its base and 50 m high. It is thus more accurate and quicker to use the equation for a pyramid.

$$\text{Vol} = \frac{AL}{3} = \frac{173.8 \times 50}{3} = 2897 \text{ m}^3$$

∴ *Total vol of fill* $= (39\,660 + 2897) = 42\,557$ m$^3$

The student should now calculate for himself the volume of cut.

(*Answer:* 39 925 m$^3$).

*Example 3.6.* The access to a tunnel has a level formation width of 10 m and runs into a plane hillside, whose natural ground slope is 1 in 10. The intersection line of this

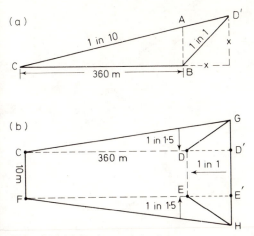

(a)

(b)

*Figure 3.25* (a) Section, and (b) plan

formation and the natural ground is perpendicular to the centre-line of the tunnel. The level formation is to run a distance of 360 m into the hillside, terminating at the base of a cutting of slope 1 vertical to 1 horizontal. The side slopes are to be 1 vertical to 1.5 horizontal.

Calculate the amount of excavation in cubic metres. Marks will be deducted if calculations are not clearly related to diagrams. (LU)

*Figure 3.25* illustrates the question which is solved by the methods previously advocated.

Height $AB = 36$ m as ground slope is 1 in 10

By rate of approach

$$x = \left(1 - \frac{1}{10}\right)^{-1} \times AB = \frac{10 \times 36}{9} = 40 \text{ m} = DD'$$

As the side slopes are 1 in 1.5 and $D'$ is 40 m high, then $D'G = 40 \times 1.5 = 60$ m $= E'H$. Therefore $GH = 130$ m

Area of $\triangle BCD'$ (in section above) normal to $GH$, $DE$, $CF$

$$= \frac{360}{2} \times 40 = 7200 \text{ m}^2$$

$$\therefore \text{ Vol} = \frac{7200}{3}(130 + 10 + 10) = 360\,000 \text{ m}^3$$

*Check*

$$\text{Wedge } BCD' = \frac{360}{6}[(130 + 10 + 10) \times 40] = 360\,000 \text{ m}^3$$

*Example 3.7.* A solid pier is to have a level top surface 20 m wide. The sides are to have a batter of 2 vertical in 1 horizontal and the seaward end is to be vertical and perpendicular to the pier axis. It is to be built on a rock stratum with a uniform slope of 1 in 24, the direction of this maximum slope making an angle whose tangent is 0.75 with

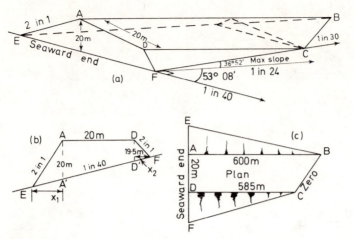

*Figure 3.26*

the direction of the pier. If the maximum height of the pier is to be 20 m above the rock, diminishing to zero at the landward end, calculate the volume of material required.                                                               (LU)

*Figure 3.26* illustrates the question. Students should note that not only is the slope in the direction of the pier required but also the slope at right-angles to the pier.

By dip and strike

tan apparent slope = tan max slope × cos included angle

$$\frac{1}{x} = \frac{1}{24} \cos 36° \ 52' \qquad \text{where } \tan^{-1} 0.75 = 36° \ 52'$$

$$x = 30$$

∴  Grade in direction of pier = 1 in 30, ∴  $AB = 20 \times 30 = 600$ m

Grade at right-angles: $\dfrac{1}{y} = \dfrac{1}{24} \cos 53° \ 08'$

$y = 40$. Grade = 1 in 40 as shown on *Figure 3.26(a)*

∴  $DD' = 19.5$ m and $DC = 19.5 \times 30 = 585$ m

From *Figure 3.26(b)*

$$x_1 = \left(2 - \frac{1}{40}\right)^{-1} \times 20 \ = \ 10.1 \text{ m}$$

$$x_2 = \left(2 + \frac{1}{40}\right)^{-1} \times 19.5 = \ 9.6 \text{ m}$$

∴  Area $\Delta EAA' = \dfrac{20 \times 10.1}{2} \qquad = 101 \text{ m}^2$

Area $\Delta DFD' = \dfrac{19.5 \times 9.6}{2} \qquad = \ 93.6 \text{ m}^2$

Now, vol $ABCD$ = plan area × mean height

$$= \left(\frac{600 + 585}{2} \times 20\right) \times \frac{1}{4}(20 + 19.5 + 0 + 0) = 117\ 315\ \text{m}^3$$

Vol of pyramid $EAB = \dfrac{\text{area } EAA' \times AB}{3} = \dfrac{101 \times 600}{3}$

$$= 20\ 200\ \text{m}^3$$

Vol of pyramid $DFC = \dfrac{\text{area } DFD' \times DC}{3} = \dfrac{93.6 \times 585}{3}$

$$= 18\ 252\ \text{m}^3$$

$Total\ vol$ = $(117\ 315 + 20\ 200 + 18\ 252) = 155\ 767\ \text{m}^3$

Alternatively, finding the area of cross-sections at chainages 0, 585/2 and 585 and applying the prismoidal rule plus treating the volume from chainage 585 to 600 as a pyramid, gives an answer of 155 525 m³.

*Example 3.8.* A 100-m length of earthwork volume for a proposed road has a constant cross-section of cut and fill, in which the cut area equals the fill area. The level formation is 30 m wide, transverse ground slope is 20° and the side slopes in cut-and-fill are $\frac{1}{2}$ horizontal to 1 vertical and 1 horizontal to 1 vertical, respectively. Calculate the volume of excavation in the 100-m length.　　　　　　　　　　　　　　(LU)

If the student turns *Figure 3.27* through 90°, then the 1-in-2.75 grade (20°) becomes 2.75 in 1 and the 2-in-1 grade becomes 1 in 2, then by rate of approach:

$$h_1 = (2.75 - 1)^{-1}(30 - x) = \frac{30 - x}{1.75}$$

$$h_2 = \left(2.75 - \frac{1}{2}\right)^{-1} x = \frac{x}{2.25}$$

Now, area $\Delta A_1 = \dfrac{30 - x}{2} \times h_1 = \dfrac{(30 - x)^2}{3.5}$

area $\Delta A_2 = \dfrac{x}{2} \times h_2 = \dfrac{x^2}{4.5}$

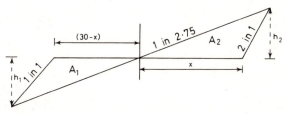

*Figure 3.27*

But area $A_1$ = area $A_2$

$$\frac{(30 - x)^2}{3.5} = \frac{x^2}{4.5}$$

$$(30 - x)^2 = \frac{3.5}{4.5}x^2 = \frac{7}{9}x^2 \qquad \text{from which } x = 16 \text{ m}$$

$$\therefore \text{ Area } A_2 = \frac{16^2}{4.5} = 56.5 \text{ m}^2 = \text{area } A_1$$

$$\therefore \text{ Vol in 100 m length} = 56.5 \times 100 = 5650 \text{ m}^3$$

*Example 3.9.* A length of existing road of formation width 20 m lies in a cutting having side slopes of 1 vertical to 2 horizontal. The centre-line of the road forms part of a circular curve having a radius of 750 m. For any cross-section along this part of the road the ground surface and formation are horizontal. At chainage 5400 m the depth to formation at the centre-line is 10 m, and at chainage 5500 m the corresponding depth is 18 m.

The formation width is to be increased by 20 m to allow for widening the carriageway and for constructing a parking area. The whole of the widening is to take place on the side of the cross-section remote from the centre of the arc, the new side slope being 1 vertical to 2 horizontal. Using the prismoidal rule, calculate the volume of excavation between the chainages 5400 m and 5500 m. Assume that the depth to formation changes uniformly with distance along the road. (ICE)

From *Figure 3.28* it can be seen that the centroid of the increased excavation lies $(20 + x)$ m from the centre-line of the curve. The distance $x$ will vary from section to section but as the side slope is 1 in 2, then:

$$x = 2 \times \frac{h}{2} = h$$

horizontal distance of centroid from centre-line = $(20 + h)$

At chainage 5400 m, $h_1 = 10$ m     $\therefore (20 + h) = 30$ m = $d_1$
At chainage 5450 m, $h_2 = 14$ m     $\therefore (20 + h) = 34$ m = $d_2$
At chainage 5500 m, $h_3 = 18$ m     $\therefore (20 + h) = 38$ m = $d_3$
Area of extra excavation at 5400 m = $10 \times 20 = 200$ m$^2$ = $A_1$
Area of extra excavation at 5450 m = $14 \times 20 = 280$ m$^2$ = $A_2$
Area of extra excavation at 5500 m = $18 \times 20 = 360$ m$^2$ = $A_3$

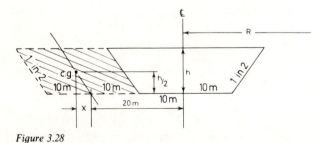

*Figure 3.28*

The above areas are now corrected for curvature: $A\left(1 + \dfrac{d}{R}\right)$

At chainage 5400 m $= 200\left(1 + \dfrac{30}{750}\right) = 208 \text{ m}^2$

At chainage 5450 m $= 280\left(1 + \dfrac{34}{750}\right) = 292.6 \text{ m}^2$

At chainage 5500 m $= 360\left(1 + \dfrac{38}{750}\right) = 378 \text{ m}^2$

$\therefore \; \text{Vol} = \dfrac{100}{6}(208 + 4 \times 292.6 + 378) = 29\,273 \text{ m}^3$

## Exercises

(*3.1*) An access road to a quarry is being cut in a plane surface in the direction of strike, the full dip of 1 in 12.86 being to the left of the direction of drive. The road is to be constructed throughout on a formation grade of 1 in 50 dipping, formation width 20 m and level, side slopes 1 in 2 and a zero depth on the centre-line at chainage 0 m.

At chainage 400 m the direction of the road turns abruptly through a clockwise angle of 40°; calculate the volume of excavation between chainages 400 m and 600 m.

(KP)

(*Answer:* 169 587 m³)

(*3.2*) A road is to be constructed on the side of a hill having a cross fall of 1 in 50 at right-angles to the centre-line of the road; the side slopes are to be 1 in 2 in cut and 1 in 3 in fill; the formation is 20 m wide and level. Find the position of the centre-line of the road with respect to the point of intersection of the formation and the natural ground.

(a) To give equality of cut and fill.
(b) So that the area of cut shall be 0.8 of the area of fill, in order to allow for bulking.

(LU)

(*Answer:* (a) 0.3 m in cut. (b) 0.2 m in fill)

(*3.3*) A reservoir is to be formed in a river valley by building a dam across it. The entire area that will be covered by the reservoir has been contoured and contours drawn at 1.5 m intervals. The lowest point in the reservoir is at a reduced level of 249 m above datum, whilst the top water level will not be above a reduced level of 264.5 m. The area enclosed by each contour and the upstream face of the dam is shown in the Table overleaf.

Estimate by the use of the trapezoidal rule the capacity of the reservoir when full. What will be the reduced level of the water surface if, in a time of drought, this volume is reduced by 25%?

(ICE)

(*Answer:* 294 211 m³; 262.3 m)

(*3.4*) The central heights of the ground above formation at three sections 100 m apart are 10 m, 12 m, 15 m, and the cross-falls at these sections are respectively 1 in 30, 1 in 40

| Contour (m) | Area enclosed (m$^2$) |
|---|---|
| 250.0 | 1 874 |
| 251.5 | 6 355 |
| 253.0 | 11 070 |
| 254.5 | 14 152 |
| 256.0 | 19 310 |
| 257.5 | 22 605 |
| 259.0 | 24 781 |
| 260.5 | 26 349 |
| 262.0 | 29 830 |
| 263.5 | 33 728 |
| 265.0 | 37 800 |

and 1 in 20. If the formation width is 40 m and sides slope 1 vertical to 2 horizontal, calculate the volume of excavation in the 200-m length:

(a) If the centre-line is straight.
(b) If the centre-line is an arc of 400 m radius.                    (LU)

(*Answer:* (a) 158 367 m$^3$. (b) 158 367 $\pm$ 1070 m$^3$)

## 3.3 MASS-HAUL DIAGRAMS

Mass-haul diagrams (MHD) are used to compare the economy of various methods of earthwork distribution on road or railway construction schemes. By the combined use of the MHD plotted directly below the longitudinal section of the survey centre-line, one can find

(1) The distances over which cut and fill will balance.
(2) Quantities of materials to be moved and the direction of movement.
(3) Areas where earth may have to be borrowed or wasted and the amounts involved.
(4) The best policy to adopt to obtain the most economic use of plant.

### 3.3.1 Definitions

(1) *Haul* refers to the volume of material multiplied by the distance moved and was expressed in 'station yards'.
(2) *Station yard* (stn yd) is 1 yd$^3$ of material moved 100 ft. At the time of writing there is no SI equivalent of this term, although it may be decided to use 1 m$^3$ moved 100 m. Thus, 20 m$^3$ moved 1500 m is a haul of $20 \times 1500/100 = 300$ stn m.
*N.B.* Hereafter the term *station metre* (stn m) will be used and defined as above.
(3) *Freehaul* and *overhaul* can best be defined by example. A contractor may offer to haul material a distance of say 150 m at 50 p per m$^3$, but thereafter for any distance hauled beyond 150 m the contractor may require an extra 5 p per stn m, i.e. 5 p per m$^3$ moved per 100 m. The distance of 150 m is called the *freehaul distance* and is based on the economical hauling distance of the earthmoving plant used. It may range from 100 m for a bulldozer to 3000 m for self-propelled scrapers. The haul beyond the freehaul distance is termed the *overhaul*.

(4) *Waste* is the material excavated from cuts but not used for embankment fills.

(5) *Borrow* is the material needed for the formation of embankments, secured not from roadway excavation but from elsewhere. It is said to be obtained from a 'borrow pit'.

(6) *Limit of economical haul* is the maximum overhaul distance plus the freehaul distance. When this limit is reached it is more economical to waste and borrow material. For example assume:

Freehaul distance = 500 m
Overhaul          = 10 p per stn m (i.e. 10 p per m³ per 100 m)
Borrow            = 30 p per m³

From these figures it can be seen that to overhaul 1 m³ a distance of 300 m would cost 30 p, equal to the cost of borrow; this then is the maximum overhaul distance. However, before overhaul comes into operation, one may move earth through the freehaul distance of 500 m. Thus the limit of economical haul = (300 + 500) = 800 m.

## 3.3.2 Bulking and shrinkage

Excavation of material causes it to loosen, thus its excavated volume will be greater than its *in situ* volume. However, when filled and compacted, it may occupy a less volume than when originally *in situ*. For example, ordinary earth is less by about 10% after filling, whilst rock bulks by some 20% to 30%. To allow for this a correction factor is generally applied to the cut or fill volumes.

## 3.3.3 Construction of the MHD

A MHD is a continuous curve, whose vertical ordinates, plotted on the same distance scale as the longitudinal section, represent the algebraic sum of the corrected volumes (cut +, fill −).

## 3.3.4 Properties of the MHD

Consider *Figure 3.29(a)* in which the ground *XYZ* is to be levelled off to the grade line *A'B'*. Assuming that the fill volumes, after correction, equal the cut volumes, the MHD would plot as shown in *Figure 3.29(b)*. Thus:

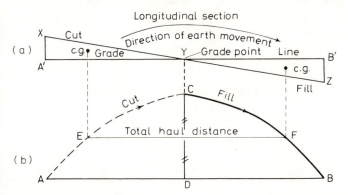

*Figure 3.29* Mass-haul diagram

(1) Since the curve of the MHD represents the algebraic sums of the volumes, then any horizontal line drawn parallel to the base *AB* will indicate the volumes which balance. Such a line is called a *balancing line* and may even be represented by *AB* itself, indicating that the total cut equals the total fill.

(2) The rising curve, shown broken, indicates cut ($+$ve), the falling curve indicates fill ($-$ve).

(3) The maximum and minimum points of a MHD occur directly beneath the intersection of the natural ground and the formation grade; such intersections are called *grade points*.

(4) As the curve of the MHD rises above the balance line *AB*, the haul is from left to right. When the curve lies below the balance line, the haul is from right to left.

(5) The total cut volume is represented by the maximum ordinate *CD*.

(6) In moving earth from cut to fill, assume the first load would be from the cut at *X* to the fill at *Y*; the last load from the cut at *Y* to the fill at *Z*. Thus the haul distance would appear to be from a point mid-way between *X* and *Y*, to a point mid-way between *Y* and *Z*. However, as the section is representative of volume not area, the haul distance is from the centroid of the cut volume to the centroid of the fill volume. The horizontal positions of these centroids may be found by bisecting the total volume ordinate *CD* with the horizontal line *EF*.

Now, since haul is volume × distance, the *total haul* in the section is total vol × total haul distance = $CD \times EF/100$ stn m.

### 3.3.5 Balancing procedures

In order to illustrate the use of freehaul distance consider *Figure 3.30.*

(1) Assuming a freehaul distance of 100 m; move this scaled distance up and down the MHD, keeping it parallel to the base *A'B'* until it cuts the curve at *E* and *F*.

(2) *EF* indicates on the longitudinal section that the cut volume *LMY* equals the fill

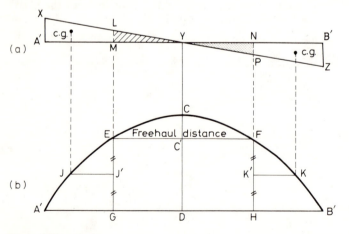

*Figure 3.30*

volume $YNP$. The amount of the volume is $CC'$ and it obviously falls within the freehaul distance.

(3) The remaining cut volume $XLMA'$ is represented by the ordinate $EG$ and is the *overhaul volume*.

(4) The overhaul volume $XLMA'$ has now to be filled into $NPZB'$, the average distance being from centroid to centroid. The positions of the centroids are found by bisecting $EG$ and $FH$, giving the horizontal distance between centroids $JK$.

(5) Assuming $JK = 250$ m, the overhaul volume has to be moved through this distance. However, the first 100 m of the movement is still within the freehaul contract, thus the overhaul distance is $(250 - 100) = 150$ m.

(6) From (5) it is obvious that the total volume $(CC' + EG) = CD$ falls within the freehaul contract.

(7) Thus, the overhaul = overhaul vol × overhaul distance = $EG(JK - EF)$.

## Worked examples

*Example 3.10.* The following notes refer to a 1200-m section of a proposed railway, and the earthwork distribution in this section is to be planned without regard to the adjoining sections. The table shows the stations and the surface levels along the centre-line, the formation level being at an elevation above datum of 43.5 m at chainage 70 and thence rising uniformly on a gradient of 1.2%. The volumes are recorded in $m^3$, the cuts are plus and fills minus.

(1) Plot the longitudinal section using a horizontal scale of 1:1200 and a vertical scale of 1:240.

(2) Assuming a correction factor of 0.8 applicable to fills plot the MHD to a vertical scale of 1000 $m^3$ to 20 mm.

(3) Calculate *total haul* in stn. m and indicate the haul limits on the curve and section.

(4) State which of the following estimates you would recommend

    (a) No freehaul at 35 p per $m^3$ for excavating, hauling and filling

    (b) A freehaul distance of 300 m at 30 p per $m^3$ plus 2 p per stn m for overhaul.

        (LU)

| Chn | Surface level | Vol | Chn | Surface level | Vol | Chn | Surface level | Vol |
|---|---|---|---|---|---|---|---|---|
| 70 | 52.8 | | 74 | 44.7 | | 78 | 49.5 | |
| | | +1860 | | | −1080 | | | −237 |
| 71 | 57.3 | | 75 | 39.7 | | 79 | 54.3 | |
| | | +1525 | | | −2025 | | | +362 |
| 72 | 53.4 | | 76 | 37.5 | | 80 | 60.9 | |
| | | + 547 | | | −2110 | | | +724 |
| 73 | 47.1 | | 77 | 41.5 | | 81 | 62.1 | |
| | | − 238 | | | −1120 | | | +430 |
| 74 | 44.7 | | 78 | 49.5 | | 82 | 78.5 | |

For answers to parts (1) and (2) see *Figure 3.31* and the values in the Table on p. 73.

72

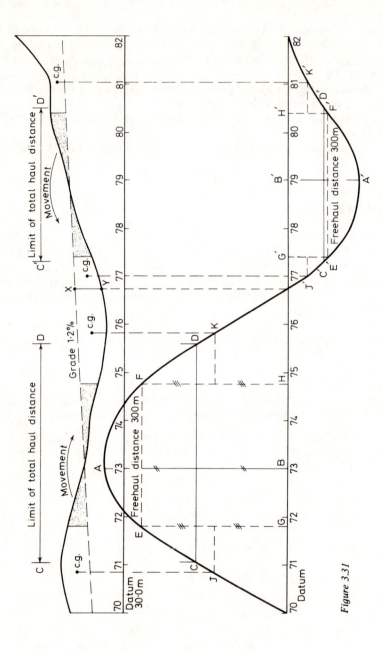

*Figure 3.31*

| Chainage | Volume | | Mass ordinate (algebraic sum) |
|---|---|---|---|
| 70 | 0 | | 0 |
| 71 | +1860 | | +1860 |
| 72 | +1525 | | +3385 |
| 73 | + 547 | | +3932 |
| 74 | − 238 × 0.8 = − | 190.4 | +3741.6 |
| 75 | − 1080 × 0.8 = − | 864 | +2877.6 |
| 76 | − 2025 × 0.8 = − 1620 | | +1257.6 |
| 77 | − 2110 × 0.8 = − 1688 | | − 430.4 |
| 78 | − 1120 × 0.8 = − | 896 | − 1326.4 |
| 79 | − 237 × 0.8 = − | 189.6 | − 1516 |
| 80 | + 362 | | − 1154 |
| 81 | + 724 | | − 430 |
| 82 | + 430 | | 0 |

N.B. (1) The volume at chainage 70 is zero.

(2) The mass ordinates are always plotted at the station and not between them.

(3) The mass ordinates are now plotted to the same horizontal scale as the longitudinal section and directly below it.

(4) Check that maximum and minimum points on the MHD are directly below grade points on the section.

(5) Using the datum line as a balancing line indicates a balancing out of the volumes from chainage 70 to $XY$ and from $XY$ to chainage 82.

Total haul (taking each loop separately) = total vol × total haul distance. The total haul distance is from the centroid of the total cut to that of the total fill and is found by bisecting $AB$ and $A'B'$, to give the distances $CD$ and $C'D'$.

$$\text{Total haul} = \frac{AB \times CD}{100} + \frac{A'B' \times C'D'}{100}$$

$$= \frac{3932 \times 450}{100} + \frac{1516 \times 320}{100} = 22\,545 \text{ stn m}$$

(a) If there is no freehaul, then all the volume is moved regardless of distance for 35 p per m$^3$.

Estimate costs: $(AB + A'B') \times 35$ p = $5448 \times 35 = 190\,680$ p

(b) The purpose of plotting the freehaul distance on the curve is to assess the overhaul.

From MHD:

Cost of freehaul = $(AB + A'B') \times 30$ p per m$^3$ (see *Section 3.3.5(6)*)

$$= 163\,440 \text{ p}$$

$$\text{Cost of overhaul} = \frac{EG(JK - EF)}{100} + \frac{E'G'(J'K' - E'F')}{100} \times 2 \text{ p}$$

$$= 13\,628 \text{ p}$$

Total cost = $163\,440 + 13\,628 = 177\,068$ p

$\therefore$ Second estimate is cheaper by 13 612 p = £136.12

N.B. All the dimensions in the above solution are scaled from the MHD.

*Example 3.11.* The volumes between sections along a 1200-m length of proposed road are shown below, positive volumes denoting cut, and negative volumes denoting fill:

| Chainage (m) | 0 | 100 | 200 | 300 | 400 | 500 | 600 | 700 | 800 | 900 | 1000 | 1100 | 1200 |
|---|---|---|---|---|---|---|---|---|---|---|---|---|---|
| Vol between sections ($m^3 \times 10^3$) | | +2.1 | +2.8 | +1.6 | −0.9 | −2.0 | −4.6 | −4.7 | −2.4 | +1.1 | +3.9 | +3.5 | +2.8 |

Plot a MHD for this length of road to a suitable scale and determine suitable positions of balancing lines so that there is

(1) A surplus at chainage 1200 but none at chainage 0.
(2) A surplus at chainage 0 but none at chainage 1200.
(3) An equal surplus at chainage 0 and chainage 1200.

Hence, determine the cost of earth removal for each of the above conditions based on the following prices and a freehaul limit of 400 m.

| | |
|---|---|
| Excavate, cart and fill (freehaul) | 60 p/m³ |
| Excavate, cart and fill (overhaul) | 85 p/m³ |
| Removal of surplus to tip from chainage 0 | 125 p/m³ |
| Removal of surplus to tip from chainage 1200 | 150 p/m³     (ICE) |

For plot of MHD see *Figure 3.32.*

Mass ordinates = 0, +2.1, +4.9, +6.5, +5.6, +3.6, −1.0,
    −5.7, −8.1, −7.0, −3.1, +0.4, +3.2 ·
    (obtained by algebraic summation of vols)

(1) Balance line *AB* gives a surplus at chainage 1200 but none at 0.
(2) Balance line *CD* gives a surplus at chainage 0 but none at 1200.
(3) Balance line *EF* situated mid-way between *AB* and *CD* will give equal surpluses at the ends.

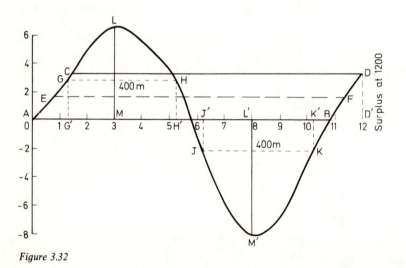

Figure 3.32

In the second part of the question the prices are quoted in an unusual manner. Excavating, carting and filling within a distance of 400 m, is at 60 p/m³. Carrying on beyond that distance gives a total price of 85 p/m³; thus the overhaul is 25 p/m³.

The question is now tackled in the usual way but it is not necessary to find overhaul distance.

(1) Taking $AB$ as base
Freehaul plots at $GH$ and $JK$.

$$\text{Cost of freehaul} = (LM + L'M') \times 60 \text{ p}$$
$$= (6500 + 8100) \times 60 = 876\,000 \text{ p}$$

brought forward    876 000 p

$$\text{Cost of overhaul} = (GG' + JJ') \times 25 \text{ p}$$
$$= (2800 + 2200) \times 25 \qquad = \quad 125\,000 \text{ p}$$

$$\text{Cost of surplus removal} = DD' \times 150 \text{ p}$$
$$= 3200 \times 150 \text{ p} \qquad\qquad = \quad 480\,000 \text{ p}$$

Total cost = 1 481 000 p

= £14 810

(2)(3) The technique is the same for these situations working to different balancing lines $CD$ and $EF$ respectively. The student should now attempt this for himself, the answers being (2) £14 130, (3) £14 380.

N.B. As the freehaul lines remain fixed, there is no overhaul on line $CD$ in the first loop of the MHD.

## Example

*Example 3.12.* Volumes in m³ of excavation ($+$) and fill ($-$) between successive sections 100 m apart on a 1300-m length of a proposed railway are given

| Section | 0 | 1 | 2 | 3 | 4 | 5 | 6 | 7 |
|---|---|---|---|---|---|---|---|---|
| Volume (m³) | | −1000 | −2200 | −1600 | −500 | +200 | +1300 | +2100 |

| Section | 7 | 8 | 9 | 10 | 11 | 12 | 13 |
|---|---|---|---|---|---|---|---|
| Volume (m³) | | +1800 | +1100 | +300 | −400 | −1200 | −1900 |

Draw a MHD for this length. If earth may be borrowed at either end, which alternative would give the least haul? Show on the diagram the forward and backward freehauls if the freehaul limit is 500 m, and give these volumes.    (LU)

Adding the volumes algebraically gives the following mass ordinates

| Section | 0 | 1 | 2 | 3 | 4 | 5 | 6 | 7 |
|---|---|---|---|---|---|---|---|---|
| Volume (m³) | | −1000 | −3200 | −4800 | −5300 | −5100 | −3800 | −1700 |

| Section | 8 | 9 | 10 | 11 | 12 | 13 |
|---|---|---|---|---|---|---|
| Volume (m³) | +100 | +1200 | +1500 | +1100 | −100 | −2000 |

These are now plotted to produce the MHD of *Figure 3.33*. Balancing out from the zero end permits borrowing at the 1300 end.

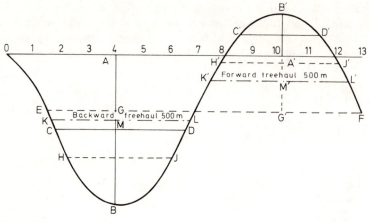

*Figure 3.33*

(1) Total haul $= \dfrac{(AB \times CD)}{100} + \dfrac{(A'B' \times C'D')}{100}$ stn m

$\phantom{(1) \text{Total haul}} = \dfrac{(5300 \times 475)}{100} + \dfrac{(1500 \times 282)}{100} = 29\ 405$ stn m

*Note:* CD bisects AB and C'D' bisects A'B'.

(2) Balancing out from the 1300 end (EF) permits borrowing at the zero end.

$\phantom{(2)} \text{Total haul} = \dfrac{(GB \times HJ)}{100} + \dfrac{(G'B' \times H'J')}{100}$

$\phantom{(2) \text{Total haul}} = \dfrac{(3300 \times 385)}{100} + \dfrac{(3500 \times 430)}{100} = 27\ 755$ stn m

Thus, borrowing at the zero end is the least alternative.

Backward freehaul $= MB\ \ = 2980$ m$^3$
Forward freehaul $\ = M'B' = 2400$ m$^3$

## 3.3.6 Auxiliary balancing lines

A study of the material on MHD plus the worked examples should have given the reader an understanding of the basics. It is now appropriate to illustrate the application of auxiliary balancing lines.

Consider in the first instance a MHD as in *Worked example 3.10*, p. 71. In *Figure 3.34* the balance line is ABC and the following data are easily extrapolated

Cut AD balances fill DB    vol moved $= DE$
Cut CJ balances fill BJ    vol moved $= HJ$

Consider now *Figure 3.35*; the balance line is AB, but in order to extrapolate the above data one requires an auxiliary balancing line CDE parallel to AB and touching the MHD at D:

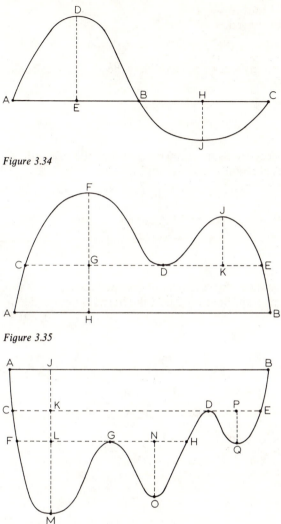

Figure 3.34

Figure 3.35

Figure 3.36

| Cut *AC* balances fill *EB* | Vol moved = *GH* |
| Cut *CF* balances fill *FD* | Vol moved = *FG* |
| Cut *DJ* balances fill *JE* | Vol moved = *JK* |

Thus the total volume moved between *A* and *B* is *FH* + *JK*

Finally, *Figure 3.36* has a balance from *A* to *B* with auxiliaries at *CDE* and *FGH*, then

| Fill *AC* balances cut *BE* | Vol moved = *JK* |
| Fill *CF* balances cut *DH* | Vol moved = *KL* |
| Fill *FM* balances cut *GM* | Vol moved = *LM* |

Fill $GO$ balances cut $HO$     Vol moved $= NO$
Fill $DQ$ balances cut $EQ$     Vol moved $= PQ$

The total volume moved between $A$ and $B$ is $JM + NO + PQ$

The above data become apparent only when one introduces the auxiliary balancing lines.

## Exercises

(*3.5*) The volumes in m$^3$ between successive sections 100 m apart on a 900-m length of a proposed road are given below (excavation is positive and fill negative):

| Section | 0 | 1 | 2 | 3 | 4 | 5 | 6 | 7 |
|---|---|---|---|---|---|---|---|---|
| Volume (m³) | | +1700 | −100 | −3200 | −3400 | −1400 | +100 | +2600 |

| Section | 7 | 8 | 9 |
|---|---|---|---|
| Volume (m³) | | +4600 | +1100 |

Determine the maximum haul distance when earth may be wasted only at the 900-m end. Show and evaluate on your diagram the overhaul if the freehaul limit is 300 m.
(LU)

(*Answer:* 558 m, 5500 stn m)

(*3.6*) Volumes of cut ($+$ve) and fill ($-$ve) along a length of proposed road are as follows:

| Chainage (m) | 0 | | 100 | | 200 | | 300 | | 400 | | 480 |
|---|---|---|---|---|---|---|---|---|---|---|---|
| Volume (m³) | | +290 | | +760 | | +1680 | | +620 | | +120 | | −20 |

| Chainage (m) | 500 | | 600 | | 700 | | 800 | | 900 | | 1000 |
|---|---|---|---|---|---|---|---|---|---|---|---|
| Volume (m³) | | −110 | | −350 | | −600 | | −780 | | −690 | | −400 |

| Chainage (m) | 1100 | | 1200 |
|---|---|---|---|
| Volume (m³) | | −120 | |

Draw a MHD and, excluding the surplus excavated material along this length, determine the overhaul if the freehaul distance is 300 m.     (ICE)

(*Answer:* 350 stn m)

**4**

# The theodolite and its application

As was shown in Chapter 1, two of the five basic measurements in plane surveying are the horizontal angle and the vertical angle. The standard instrument used in the measurement of these values is the *theodolite*.

Although there is a large variety of theodolites produced by various manufacturers, the instruments all have the same basic components arranged in the same geometric relationship. Thus, mastery of one particular instrument enables an operator to master easily any other make encountered on site.

Theodolites are normally classified by the precision to which they resolve angles and vary from 1 minute of arc (1′) to 0.1 second (0.1″), depending upon accuracy requirements of the work in hand. In engineering survey, the finesse of selecting a particular instrument appropriate to the accuracy of the survey is no longer the practice. For obvious commercial reasons, a 1″ theodolite is used for all surveys, even vertical-staff tacheometry, regardless of the accuracy requirements.

It is not intended here to deal with the practical aspects of setting up, centering over a survey station and carrying out the observations. These procedures can be properly learned only by practical work in the laboratory or field. However, the procedures necessary to maintain the instrument in good adjustment and the errors that result from maladjustment, are dealt with here in detail; along with the application of the theodolite in establishing horizontal control networks by the method of traversing.

## 4.1 BASIC CONCEPTS

There are basically two types of theodolite, (a) the vernier type, and (b) the modern glass-arc type which has largely superseded the vernier model. The terms used describe their mode of construction: in (a) the manner of reading the horizontal and vertical scales is by means of a vernier; in (b) the horizontal and vertical circles are made of glass on which the divisions have been photographically etched.

Although the vernier type is virtually obsolete, it is used in *Figure 4.1* to illustrate the basic features of all types of theodolite.

(1) The *trivet stage*, forming the base of the instrument, connects the theodolite to the tripod.
(2) The *tribrach* supports the rest of the instrument and, with reference to the *plate bubble*, can be levelled using the *footscrews* acting against the fixed trivet stage.

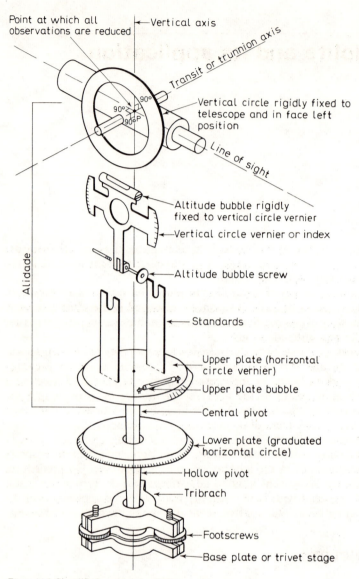

Point at which all observations are reduced

Vertical axis

Transit or trunnion axis

Vertical circle rigidly fixed to telescope and in face left position

Line of sight

Altitude bubble rigidly fixed to vertical circle vernier

Vertical circle vernier or index

Altitude bubble screw

Standards

Upper plate (horizontal circle vernier)

Upper plate bubble

Central pivot

Lower plate (graduated horizontal circle)

Hollow pivot

Tribrach

Footscrews

Base plate or trivet stage

Alidade

*Figure 4.1* Simplified vernier theodolite

(3) The *lower plate* is graduated clockwise from 0° to 360° and may be regarded as a simple circular protractor for measuring horizontal angles.

(4) The *upper plate* fits concentric with the lower plate and may be regarded as the index against which the lower plate is read. Even though in modern theodolites the lower plate is a glass arc read via a complicated optical train of lenses and prisms, the above simple concept still holds true.

Thus, to measure the horizontal angle *ABC* in *Figure 4.2* the theodolite is set up, levelled and carefully centred over Station B. The telescope is backsighted (BS) to bisect the target at A and the horizontal circle reading (20°) noted. With the lower

A

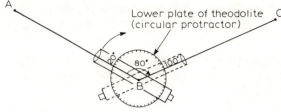

Lower plate of theodolite C
(circular protractor)

*Figure 4.2*

plate held fixed, the upper plate is rotated to $C$, the foresight (FS) target carefully bisected and the reading (100°) noted. Thus, the measured angle is (100° − 20°) = 80°. Similarly, a BS reading to $A$ of, say, 350° and a FS reading to $C$ of 70°, produces an angle of (70° − 350°) = 80°. This is the basic concept of measuring horizontal angles with a theodolite.

(5) The *standards* support the *vertical circle* and *telescope* by the *transit axis*. The standards must be high enough to permit rotation of the telescope in the vertical plane.

(6) The vertical circle (used in the measurement of vertical angles) is rigidly fixed to, and rotates with, the telescope.

(7) The *vertical circle vernier* or *index* remains fixed relative to the vertical circle and is the datum from which vertical angles are measured.

(8) The *altitude bubble* is attached to the vertical circle index and when centralized establishes the vertical circle index horizontal. Thus vertical angles must only be read off the vertical circle when the index has been set to form a horizontal datum (*Figure 4.3*). In this instance the zenith of the vertical circle reads 0°, thus a reading of, say, 120° as shown, represents a vertical angle ($\alpha$) of +30°.

Many theodolites possess automatic vertical circle indexing which does not require the setting of an altitude bubble, thereby eliminating the possibility of forgetting to do so.

## 4.2 TESTS AND ADJUSTMENTS

To ensure the best possible results when observing, the primary axes of the instrument should have the following relationship (as illustrated in *Figure 4.1*):

(a) The vertical axis of the instrument should be truly vertical when the plate level bubble is central.

(b) The transit axis should be perpendicular to the line of sight in the horizontal plane and the instrument axis in the vertical plane.

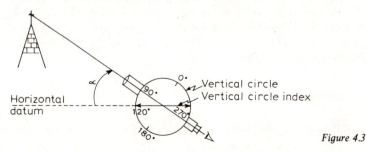

Horizontal
datum

Vertical circle
Vertical circle index

*Figure 4.3*

(c) When the telescope is horizontal and the altitude bubble central, the vertical circle should read zero or the equivalent (depending upon the manner of graduating the circle).

In order to establish the above relationships, the instrument should be tested and, if necessary, adjusted at regular intervals. The procedures for these are now given in the order in which they should be carried out.

## (1) Plate level test

The purpose of this test is stated in *Section 4.2(a)*. The vertical axis of the instrument is perpendicular to the horizontal plate which carries the plate bubble. Thus to ensure that the vertical axis of the instrument is truly vertical, as defined by the bubble, it is necessary to align the bubble axis parallel to the horizontal plate.

*Test:* Assume the bubble is not parallel to the horizontal plate but is in error by angle $e$. It is set parallel to a pair of footscrews, levelled approximately, then turned through 90° and levelled again using the third footscrew only. It is now returned to its former position, accurately levelled using the pair of footscrews, and will appear as in *Figure 4.4(a)*. The instrument is now turned through 180° and will appear as in *Figure 4.4(b)*, i.e. the bubble will move off centre by an amount representing twice the error in the instrument ($2e$).

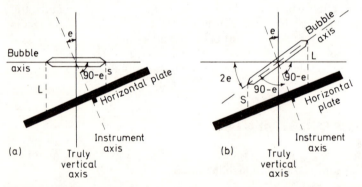

*Figure 4.4* (a) When levelled over two footscrews. (b) When turned through 180°

*Adjustment:* The bubble is brought half-way back to the centre using the pair of footscrews. This will cause the instrument axis to move through $e$ thereby making it truly vertical and, in the event of there being no adjusting tools available, the instrument may be used at this stage. The bubble will still be off centre by an amount proportional to $e$, and should now be centralized by raising or lowering one end of the bubble using its capstan adjusting screws.

## (2) Collimation in azimuth

The purpose of this test is to ensure that the line of sight is perpendicular to the transit axis.

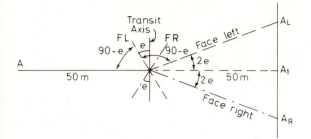

*Figure 4.5* Collimation in azimuth

*Test :* The instrument is set up, levelled, and the telescope directed to bisect a fine mark at $A$, situated at instrument height about 50 m away (*Figure 4.5*). If the line of sight is perpendicular to the transit axis, then when the telescope is rotated vertically through 180°, it will intersect at $A_1$. However, assume that the line of sight makes an angle of $(90° - e)$ with the transit axis, as shown dotted in the face left (FL) and face right (FR) positions. Then in the FL position the instrument would establish a fine mark at $A_L$. Change face, re-bisect point $A$, transit the telescope and establish a fine mark at $A_R$. From the sketch it is obvious that distance $A_L A_R$ represents four times the error in the instrument (4$e$). (Looking through the telescope of the theodolite with the vertical circle on the left is termed a *face left observation*, and *vice versa*.)

*Adjustment :* The cross-hairs are now moved in azimuth using their horizontal capstan adjusting screws, from $A_R$ to a point mid-way between $A_R$ and $A_1$; this is one-quarter of the distance $A_L A_R$.

This movement of the reticule carrying the cross-hair may cause the position of the vertical hair to be disturbed in relation to the transit axis; i.e. it should be perpendicular to the transit axis. It can be tested by traversing the telescope vertically over a fine dot. If the vertical cross-hair moves off the dot then it is not at right angles to the transit axis and is corrected with the adjusting screws.

This test is frequently referred to as one which ensures the verticality of the vertical hair, which will be true only if the transit axis is truly horizontal. However, it can be carried out when the theodolite is not levelled, and it is for this reason that a dot should be used and not a plumb line as is sometimes advocated.

## (3)  Spire test (transit axis test)

This test ensures that the transit axis is perpendicular to the vertical axis of the instrument.

*Test :* The instrument is set up and carefully levelled approximately 50 m from a well-defined point of high elevation, preferably greater than 30° (*Figure 4.6*). The well-defined point $A$ is bisected and the telescope then lowered to its horizontal position and a further point made. If the transit axis is in adjustment the point will appear at $A_1$ directly below $A$. If, however, it is in error by the amount $e$ (transit axis shown dotted in FL and FR positions), the mark will be made at $A_L$. The instrument is now changed to FR, point $A$ bisected again and the telescope lowered to the horizontal, to fix point $A_R$. The distance $A_L A_R$ is twice the error in the instrument (2$e$).

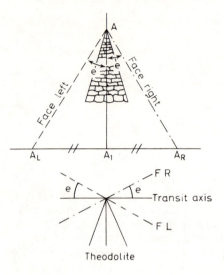

*Figure 4.6* Spire test (transit axis test)

*Adjustment:* Length $A_L A_R$ is bisected and a fine mark made at $A_1$. The instrument is now moved in azimuth, using a plate tangent screw until $A_1$ is bisected. The student should note that no adjustment of any kind has yet been made to the instrument. Thus, when the telescope is raised back to $A$ it will be in error by the horizontal distance $A_L A_R/2$. By moving one end of the transit axis using the adjusting screws, the line of sight is made to bisect $A$. This can only be made to bisect $A$ when the line of sight is elevated. Movement of the transit axis when the telescope is in the horizontal plane $A_L A_R$, will not move the line of sight to $A_1$, hence the need to incline steeply the line of sight.

## (4)  Vertical circle index test

To ensure that when the telescope is horizontal and the altitude bubble central, the vertical circle reads zero (or its equivalent).

*Test:* Centralize the altitude bubble using the clip screw (altitude bubble screw) and, by rotating the telescope, set the vertical circle to read zero (or its equivalent for a horizontal sight).

Note the reading on a vertical staff held about 50 m away. Change face and repeat the whole procedure. If error is present, a different reading on each face is obtained, namely $A_L$ and $A_R$ in *Figure 4.7*.

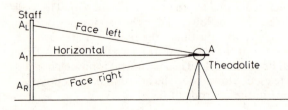

*Figure 4.7* Vertical circle index test

*Adjustment:* Set the telescope to read the mean of the above two readings, thus making it truly horizontal. The vertical circle will then no longer read zero, and must be brought back to zero without affecting the horizontal level of the telescope. This is done by moving the verniers to read zero using the clip screw.

Movement of the clip screw will cause the altitude bubble to move off centre. It is re-centralized by means of its capstan adjusting screws.

### 4.2.1 Alternative approach

#### (1) Plate level test

The procedure for this is as already described in *Section 4.2(1)*.

#### (2) Collimation in azimuth

With the telescope horizontal and the instrument carefully levelled, sight a fine mark and note the reading. Change face and repeat the procedure. If the instrument is in adjustment, the two readings should differ by exactly 180°. If not, the instrument is set to the corrected reading as shown below using the upper plate slow-motion screw; the line of sight is brought back on to the fine mark by adjusting the cross-hairs.

e.g.  FL reading        01° 30′ 20″
      FR reading        181° 31′ 40″
      ─────────────────────────────
      Difference $= 2e =$       01′ 20″
      $\therefore$        $e = \pm 40''$

      Corrected reading $= 181° 31′ 00″$ or $01° 31′ 00″$

#### (3) Spire test

With instrument carefully levelled, sight a fine point of high elevation and note the horizontal circle reading. Change face and repeat. If error is present, set the horizontal circle to the corrected reading, as above. Adjust the line of sight back on to the mark by raising or lowering the transit axis. It is worth noting here that not all modern instruments are capable of this adjustment.

#### (4) Vertical circle index test

Assume the instrument reads 0° on the vertical circle when the telescope is horizontal and in FL position. Carefully level the instrument, horizontalize the altitude bubble and sight a fine point of high elevation. Change face and repeat. The two vertical circle readings should *sum* to 180°, any difference being twice the index error.

e.g.  FL reading (*Figure 4.8(a)*)  09° 58′ 00″
      FR reading (*Figure 4.8(b)*) 170° 00′ 20″
      ─────────────────────────────────────
                    sum $= 179° 58′ 20″$
            correct sum $= 180° 00′ 00″$
      ─────────────────────────────────────
                    $2e =$       $-01′ 40″$
                    $e =$          $-50″$

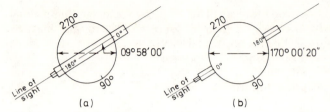

*Figure 4.8* (a) Face left, and (b) face right

Thus with target still bisected, the vernier is set to read $170° 00' 20'' + 50'' = 170° 01' 10''$ by means of the clip or altitude bubble screw. The altitude bubble is then centralized using its capstan adjusting screws. If the vertical circle reads 90° and 270° instead of 0° and 180°, the readings sum to 360°.

These alternative procedures have the great advantage of using the theodolite's own scales, rather than external scales, and can therefore be carried out by one person.

### 4.2.2 Effect of instrument errors

Adjustments are never perfect, small residual errors always remain in the instrument. Their effect will now be considered in detail. A careful study of this section will clearly show the reasons for always observing angles more than once, on alternate faces of the instrument.

### (1) Eccentricity of centres

This error is due to the centre of the central pivot carrying the alidade (upper part of the instrument) not coinciding with the centre of the hollow pivot carrying the graduated circle (*Figures 4.1* and *4.9*).

The effect of this error on readings is periodic. If $B$ is the centre of the graduated circle and $A$ is the centre about which the alidade revolves, then distance $AB$ is interpreted as an arc $ab$ in seconds on the graduated circle and is called the *error of eccentricity*. If a vernier is at $D$, on the line of the two centres, it reads the same as it would if there were no error. If, at $b$, it is in error by $ba = E$, the maximum error. In an intermediate position $d$, the error will be $de = BC = AB \sin \theta = E \sin \theta$, $\theta$ being the horizontal angle of rotation.

The horizontal circle is graduated clockwise, thus the vernier supposedly at $b$ will be at $a$, giving a reading too great by $+E$. The opposite vernier supposedly at $b'$ will be at

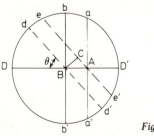

*Figure 4.9*

$a'$, thereby reading too small by $-E$. Similarly for the intermediate positions at $d$ and $d'$, the errors will be $+E\sin\theta$ and $-E\sin\theta$. *Thus the mean of the two verniers 180° apart, will be free of error.*

Modern glass-arc instruments in the 20″ class, can be read on one side of the graduated circle only, thus producing an error which varies sinusoidally with the angle of rotation. Readings on both faces of the instrument would establish verniers 180° apart. *Thus the mean of readings on both faces of the instrument will be free of error.* With 1-in theodolites the readings 180° apart on the circle are automatically averaged and so are free of this error.

Manufacturers claim that this source of error does not arise in the construction of modern glass-arc instruments.

## (2)  Collimation in azimuth error

If the line of sight in *Figure 4.10* is at right-angles to the transit axis it will sweep out the vertical plane $VOA$ when the telescope is depressed through the vertical angle $\alpha$.

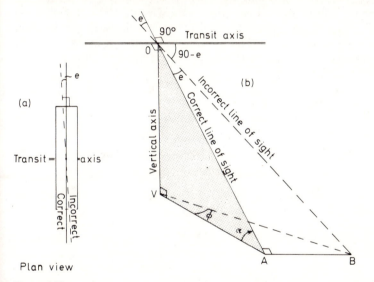

Figure 4.10

If the line of sight is not at right-angles but in error by an amount $e$, the vertical plane swept out will be $VOB$. Thus the pointing is in error by $-\phi$ ($-$ve because the horizontal circle is graduated clockwise).

$$\tan\phi = \frac{AB}{VA} = \frac{OA\tan e}{VA} \qquad \text{but} \qquad \frac{OA}{VA} = \sec\alpha$$

$$\therefore\ \tan\phi = \sec\alpha\tan e$$

as $\phi$ and $e$ are very small, the above may be written

$$\phi = e\sec\alpha \tag{4.1}$$

On changing face $VOB$ will fall to the other side of $A$ and give an equal error of opposite sign, i.e. $+\phi$. *Thus the mean of readings on both faces of the instrument will be free of error.*

$\phi$ is the error of one sighting to a target of elevation $\alpha$. An angle, however, is the difference between two sightings; therefore the error in an angle between two objects of elevation, $\alpha_1$ and $\alpha_2$, will be $e(\sec \alpha_1 - \sec \alpha_2)$ and will obviously be zero if $\alpha_1 = \alpha_2$, or if measured in the horizontal plane, $(\alpha = 0)$.

On the opposite face the error in the angle simply changes sign to $-e(\sec \alpha_1 - \sec \alpha_2)$, indicating that the *mean of the two angles taken on each face will be free of error regardless of elevation*.

*Vertical angles*: It can be illustrated that the error in the measurement of the vertical angles is $\sin \alpha = \sin \alpha_1 \cos e$ where $\alpha$ is the measured altitude and $\alpha_1$ the true altitude. However, as $e$ is very cmall, $\cos e \approx 1$, hence $\alpha_1 \approx \alpha$, proving that the effect of this error on vertical angles is negligible.

## (3)  Effect of transit axis dislevelment

If the transit axis is set correctly at right-angles to the vertical axis, then when the telescope is depressed it will sweep out the truly vertical plane $VOA$ (*Figure 4.11*). Assuming the transit axis is inclined to the horizontal by $e$, it will sweep out the plane $COB$ which is inclined to the vertical by $e$. This will create an error $-\phi$ in the horizontal reading of the theodolite ($-$ve as the horizontal circle is graduated clockwise).

If $\alpha$ is the angle of inclination then

$$\sin \phi = \frac{AB}{VB} = \frac{VC}{VB} = \frac{OV}{VB} \tan e = \tan \alpha \tan e$$

Now, as $\phi$ and $e$ are small, $\phi = e \tan \alpha$ \hfill (4.2)

From *Figure 4.11* it can be seen that the correction $\phi$ to the reading at $B$, to give the

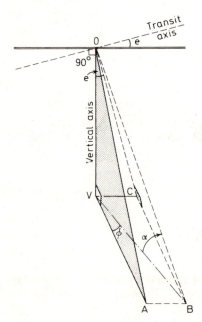

Figure 4.11

correct reading at $A$, is positive because of the clockwise graduations of the horizontal circle. Thus, when looking through the telescope towards the object, if the left-hand end of the transit axis is high, then the correction to the reading is positive, and *vice versa*.

On changing face, $COB$ will fall to the other side of $A$ and give an equal error of opposite sign. *Thus, the mean of the readings on both faces of the instrument will be free of error.*

As previously, the error in the measurement of an angle between two objects of elevations $\alpha_1$ and $\alpha_2$ will be

$$e(\tan \alpha_1 - \tan \alpha_2)$$

which on changing face becomes $-e(\tan \alpha_1 - \tan \alpha_2)$ indicating *that the mean of two angles, taken one on each face, will be free of error regardless of elevation.* Also if $\alpha_1 = \alpha_2$, or the angle is measured in the horizontal plane ($\alpha = 0$), it will be free of error. Note if $\alpha_1$ is positive and $\alpha_2$ negative, then the correction is $e[\tan \alpha_1 - (-\tan \alpha_2)] = e(\tan \alpha_1 + \tan \alpha_2)$.

*Vertical angles:* Errors in the measurement of vertical angles can be shown to be $\sin \alpha = \sin \alpha_1 \sec e$. As $e$ is very small $\sec e \approx 1$, thus $\alpha_1 = \alpha$, proving that the effect of this error on vertical angles is negligible.

## (4)  Effect of non-verticality of the vertical axis

If the plate levels of the theodolite are not in adjustment, then the instrument axis will be inclined to the vertical, hence measured azimuth angles will not be truly horizontal. Assuming the transit axis is in adjustment, i.e. perpendicular to the vertical axis, then error in the vertical axis of $e$ will cause the transit axis to be inclined to the horizontal by $e$, producing an error in pointing of $\phi = e \tan \alpha$ as in the previous case. Here, however, the error is not eliminated by double-face observations (*Figure 4.12*), but varies on different pointings of the telescope. For example, *Figure 4.13(a)* shows the instrument axis truly vertical and the transit axis truly horizontal. Imagine now that the instrument axis is inclined through $e$ in a plane at 90° to the plane of the paper (*Figure 4.13(b)*). There is no error in the transit axis. If the alidade is now rotated clockwise through 90° into the plane of the paper, it will be as in *Figure 4.13(c)*, and when viewed in the direction of the arrow, will appear as in *Figure 4.13(d)* with the transit axis inclined to the horizontal by the same amount as the vertical axis, $e$. Thus, the error in the transit axis varies from zero to maximum through 90°. At 180° it will be zero again, and at 270° back to maximum in exactly the same position.

If the horizontal angle between the plane of the transit axis and the plane of dislevelment of the vertical axis is $\delta$, then the transit axis will be inclined to the horizontal by $e \cos \delta$. For example, in *Figure 4.13(b)*, $\delta = 90°$, therefore as $\cos 90° = 0$, the inclination of the transit axis is zero, as shown.

For an angle between two targets at elevations $\alpha_1$ and $\alpha_2$, in directions $\delta_1$ and $\delta_2$, the correction will be $e(\cos \delta_1 \tan \alpha_1 - \cos \delta_2 \tan \alpha_2)$. When $\delta_1 = \delta_2$, the correction is a maximum when $\alpha_1$ and $\alpha_2$ have opposite signs. When $\delta_1 = -\delta_2$, that is in opposite directions, the correction is maximum when $\alpha_1$ and $\alpha_2$ have the same sign.

If the instrument axis is inclined to the vertical by an amount $e$ and the transit axis further inclined to the horizontal by an amount $i$, both in the same plane, then the maximum dislevelment of the transit axis on one face will be $(e + i)$, and $(e - i)$ on the

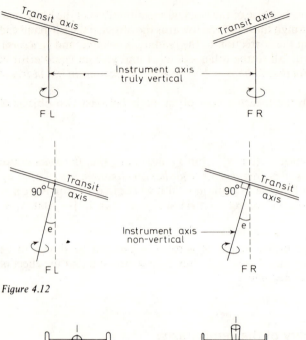

*Figure 4.12*

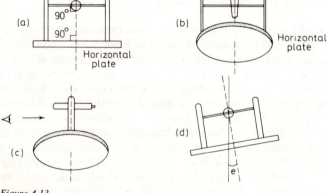

*Figure 4.13*

reverse face (*Figure 4.14*). Thus, the correction to a pointing on one face will be $(e + i) \tan \alpha$ and on the other $(e - i) \tan \alpha$, resulting in a correction of $e \tan \alpha$ to the mean of both face readings.

As shown, the resultant error increases as the angle of elevation $\alpha$ increases and is not eliminated by double-face observations. As steep sights frequently occur in mining and civil engineering surveys, it is very important to recognize this source of error and adopt the correct procedures.

Thus, as already illustrated, the correction for a specific direction $\delta$ due to non-verticality $(e)$ of the instrument axis is $e \cos \delta \tan \alpha$. The value of $e \cos \delta = E$ can be obtained from

$$E'' = S'' \frac{(L - R)}{2} \tag{4.3}$$

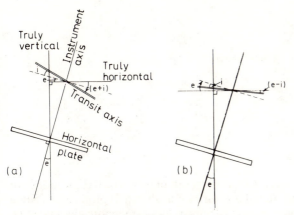

*Figure 4.14* (a) Face left, and (b) face right

where   $S''$ = the sensitivity of the plate bubble in secs of arc per bubble division
$L$ and $R$ = the left- and right-hand readings of the ends of the plate bubble when viewed from the eye-piece end of the telescope

Then the correction to each horizontal circle reading is $C'' = E'' \tan \alpha$, and is plus when $L > R$ and *vice versa*.

For high-accuracy survey work, the accuracy of the correction $C$ will depend upon how accurately $E$ can be assessed. This, in turn, will depend on the sensitivity of the plate bubble and how accurately the ends can be read. For very high accuracy involving extremely steep sights, an Electrolevel attached to the theodolite will measure axis tilt direct. This instrument has a sensitivity of 1 scale division equal to $1''$ of tilt and can be read to 0.25 div. The average plate bubble has a sensitivity of $20''$ per div.

Assuming that one can read each end of the plate bubble to an accuracy of $\pm 0.5$ mm, then for a bubble sensitivity of $20''$ per (2 mm) div, on a vertical angle of $45°$, the error in levelling the instrument (i.e. in the vertical axis) would be $\pm 0.35 \times 20'' \tan 45° = \pm 7''$. It has been shown that the accuracy of reading a bubble through a split-image coincidence system is ten times greater. Thus, if the altitude bubble, usually viewed through a coincidence system, was used to level the theodolite, error in the axis tilt would be reduced to $\pm 0.7''$.

Modern theodolites are rapidly replacing the altitude bubble with automatic vertical circle indexing with stabilization accuracies of $\pm 0.3''$, which may therefore be used for high-accuracy levelling of the instrument as follows:

(1) Accurately level the instrument using its plate bubble in the normal way.
(2) Clamp the telescope in any position, place the plane of the vertical circle parallel to two footscrews and note the vertical circle reading.
(3) With telescope remaining clamped, rotate the alidade through $180°$ and note vertical circle reading.
(4) Using the two footscrews of (2) above, set the vertical circle to the mean of the two readings obtained in (2) and (3).
(5) Rotate through $90°$ and by using only the remaining footscrew obtain the same mean vertical circle reading.

The instrument is now precisely levelled to minimized axis tilt and virtually eliminates this source of error on steep sights.

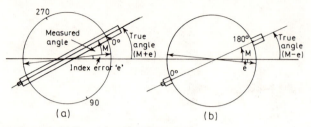

*Figure 4.15* (a) Face left, and (b) face right

## (5)  Vertical circle index error

This form of error (illustrated in *Figure 4.15*) is the result of the vertical circle index not being exactly horizontal but having an index error of '*e*' as shown. It produces an equal but opposite error on alternate faces of the instrument and is therefore eliminated by taking the mean of double-face readings.

## (6)  Plate graduation errors

Periodic errors in the graduation of the horizontal and vertical circles of a theodolite occur due to the manufacturing processes involved. The production of error curves for some instruments indicated maximum errors in the region of $\pm 0.3''$. Thus, for all practical considerations, reading on different parts of the circle and taking the mean, should be sufficient treatment for this particular error.

Plate graduation errors combined with optical and mechanical defects in the micrometer system produced optical micrometer errors which exhibit a cyclic variation. Their effect can be reduced by using different micrometer settings.

It is obvious from a consideration of all these error sources, that not only should the observations be made several times on alternate faces of the instrument, but that the instrument should be handled with extreme care and delicacy.

A study of the following booking form will reveal the procedure adopted and the manner in which perambulations about the instrument are reduced to a minimum, when observing horizontal angle *ABC* (see *Figure 4.2*).

| Sight to | Face | Reading | | | Angle | | |
|----------|------|---|---|---|---|---|---|
| | | ° | ′ | ″ | ° | ′ | ″ |
| A | L | 020 | 46 | 28 | 80 | 12 | 06 |
| C | L | 100 | 58 | 34 | | | |
| C | R | 280 | 58 | 32 | 80 | 12 | 08 |
| A | R | 200 | 46 | 24 | | | |
| A | R | 292 | 10 | 21 | 80 | 12 | 07 |
| C | R | 012 | 22 | 28 | | | |
| C | L | 192 | 22 | 23 | 80 | 12 | 04 |
| A | L | 112 | 10 | 19 | | | |
| | | | | Mean = | 80 | 12 | 06 |

Note the built-in checks supplied by changing face, i.e. the reading should change by 180°.
Note that to obtain the clockwise angle one always deducts BS (*A*) reading from the FS (*C*) reading, regardless of the order in which they are observed.

## Examples

No answers are given to these questions as it would be a repetition of information already supplied. Students are advised to write out the answers for themselves.

*Example 1.* A modern theodolite is offered for sale on seven days' approval. Superficially it appears to be in good condition. List the tests you would carry out to determine whether or not the instrument was fit for immediate use. Against each test indicate the defects that would be revealed and state whether these can be corrected by either observation techniques or field adjustments, or only by the manufacturer. Also describe in detail the procedure involved in making two of the field adjustments in your list.
*Note:* Field adjustments are those for which the manufacturer normally provides a tool kit in the instrument case. (LU)

*Example 2.* The trunnion axis of a theodolite is set at an angle of $(90° - i)$ to the vertical axis of the instrument, where $i$ is small. Derive an expression for the error in measuring the horizontal angle subtended by two objects whose angles of elevation are $\alpha$ and $\beta$.

The trunnion axis of a certain theodolite is 80 mm long and has one end 0.005 mm higher than the other. Evaluate the error in the observed azimuth angle between a star at an elevation of 65° and a reference object at a depression of 5° if the observations are made on one face only. (LU)

*Example 3.* Show that the effect of an eccentricity of the horizontal circle of a theodolite is to produce an error in reading one side only, which varies sinusoidally with the angle of rotation of the telescope. Show also that, where only one side of the circle may be read, provided that the azimuth of the circle remains unchanged, the mean of the FL and FR readings gives the correct angle. (LU)

*Example 4.* Two stations at elevations of $\alpha_1$ and $\alpha_2$ are sighted by a theodolite in which the line of collimation is inclined to the trunnion axis at an angle of $(90° - i)$ where $i$ is small.

(1) Derive an expression for the error in the horizontal angle between the two stations, as given by this instrument.
(2) Show by a diagram the effect of the collimation error on the vertical circle reading to one station.
(3) What is the effect of measuring the horizontal and vertical angles on both faces? (LU)

## 4.3 THEODOLITE TRAVERSING

Horizontal control networks (as outlined in Chapter 1) can be established by a variety of methods, the method under consideration here is that of *traversing*. The positions of the control points (survey stations) are fixed by measuring the *horizontal angles*, at each station, subtended by the adjacent stations and the *horizontal distance* between consecutive pairs of stations, as shown in *Figure 4.16*. The angles define the shape of the network, whilst the lengths establish the scale.

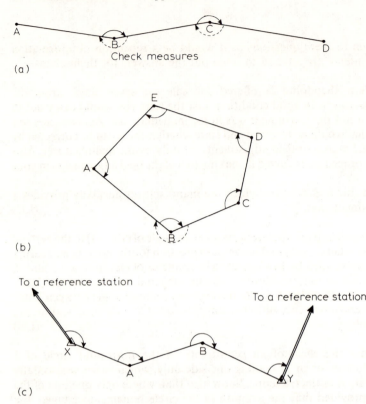

Figure 4.16 Types of traverse: (a) open, (b) polygonal and (c) link

In engineering, traverse networks are used:

(1) As control for topographic detail surveying.
(2) As control for dimensional control (setting out) on site.
(3) As control for aerial surveys.

### 4.3.1 Types of traverse

#### (1) Open traverse

This type of traverse (*Figure 4.16(a)*) neither returns to its starting point nor connects in to any other known point. Because of this, there is no automatic check on the field data, and open traverses should therefore be avoided where possible.

Open traverses are used mainly in tunnelling work where the physical situation prevents closure. It is important therefore that the measured angles, distances and instrument centring, be very carefully checked by independent means, wherever possible.

## (2) Closed traverse

A *closed polygonal traverse* is one which closes back on to its starting point (*Figure 4.16(b)*).

A closed traverse which commences from a point (*X*) of known value and connects into a second point (*Y*) of known value, is called a *link traverse* (*Figure 4.16(c)*).

The advantage of a closed traverse is that the amount of angular and linear misclosure can be detected and so distributed throughout the traverse, thereby rendering it geometrically correct.

It is worth noting that if the sides of a polygonal traverse are measured with a tape which is too long or too short, the polygon will be too small or too large, but will appear to close satisfactorily. Also, if the first line of the traverse is in error by, say, $\theta°$, the whole network will swing through $\theta°$ but will still appear to close satisfactorily. Both types of gross error will be immediately apparent in a link traverse.

## 4.4 TRAVERSE COMPUTATION

### 4.4.1 Distribution of angular error

On the measurement of the horizontal angles of the traverse, the majority of the systematic errors are eliminated by repeated double-face observation. The remaining random errors are distributed equally around the network, as follows.

In a polygon the sum of the *internal* angles should equal $(2n - 4)90°$, the sum of the *external* angles should equal $(2n + 4)90°$.

$$\therefore \text{ Angular misclosure} = W = \sum_{i=1}^{n} \alpha_i - (2n \pm 4)90°$$

where   $\alpha$ = mean observed angle
$n$ = number of angles in the traverse

The angular misclosure $W$ is now distributed by equal amounts on each angle, thus:

Correction per angle = $W/n$ (refer *Table 4.1*, p. 96)

However, before the angles are corrected, the angular misclosure $W$ must be considered to be acceptable. If $W$ was too great, and therefore indicative of poor observations, the whole traverse may need to be re-measured. A method of assessing the acceptability or otherwise of $W$ is given in *Section 4.4.2*.

### 4.4.2 Acceptable angular misclosure

The following procedure may be adopted provided that there is evidence of the variance of the mean observed angles, i.e.

$$\sigma_w^2 = \sigma_{a1}^2 + \sigma_{a2}^2 + \ldots + \sigma_{an}^2$$

where   $\sigma_{an}^2$ = variance of the mean observed angle

$\sigma_w^2$ = variance of the sum of the angles of the traverse

**TABLE 4.1**

| Angle | Observed value ° ′ ″ | Corrn ″ | Corrected angle ° ′ ″ | WCB ° ′ ″ | Line |
|---|---|---|---|---|---|
| | | | | 0 00 00 | AB |
| ABC | 120 20 00 | +5 | 120 20 05 | 300 20 05 | BC |
| BCD | 86 00 40 | +5 | 86 00 45 | 206 20 50 | CD |
| CDE | 341 34 20 | +5 | 341 34 25 | 07 55 15 | DE |
| DEF | 60 22 00 | +5 | 60 22 05 | 248 17 20 | EF |
| EFA | 100 28 20 | +5 | 100 28 25 | 168 45 45 | FA |
| FAB | 11 14 10 | +5 | 11 14 15 | 0 00 00 | AB |
| | 719 59 30 | +30 | 720 00 00 | check | |

$(2n - 4)90° = 720\ \ 00\ \ 00$

Error $= -30$    Correction per angle $= \dfrac{+30''}{6} = +5''$

Assuming that each angle is measured with equal precision:

$$\sigma_{a1}^2 = \sigma_{a2}^2 = \ldots \sigma_{an}^2 = \sigma_A^2$$

then $\sigma_w^2 = n . \sigma_A^2$ and

$$\sigma_w = n^{1/2} . \sigma_A \quad \text{(see Chapter 1, Volume 2)}$$

Angular misclosure $= W = \displaystyle\sum_{i=1}^{n} \alpha_i - \left[ (2n \pm 4)90° \right]$

where $\alpha$ = mean observed angle

$n$ = number of angles in traverse

then for 95% confidence

$$P(-1.96\sigma_w < W < +1.96\sigma_w) = 0.95$$

and for 99.73% confidence

$$P(-3\sigma_w < W < +3\sigma_w) = 0.9973$$

e.g. Consider a closed traverse of nine angles. Tests prior to the survey showed that the particular theodolite used had a standard error ($\sigma_A$) of $\pm 3''$. What would be considered an acceptable angular misclosure for the traverse?

$$\sigma_w = 9^{1/2} . 3'' = \pm 9''$$

$$P(-1.96 \times 9'' < W < +1.96 \times 9'') = 0.95$$

$$= P(-18'' < W < +18'') = 0.95$$

Similarly $P(-27'' < W < +27'') = 0.9973$

Thus, if the angular misclosure $W$ is greater than $\pm 18''$ there is evidence to suggest unacceptable error in the observed angles, provided the estimate for $\sigma_A$ is reliable. If $W$ exceeds $\pm 27''$ there is definitely angular error present of such proportions as to be quite unacceptable.

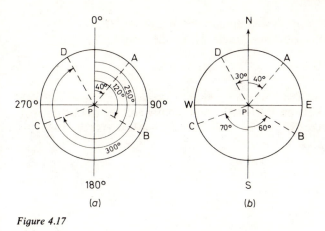

*Figure 4.17*

Research has shown that a reasonable value for the standard error of a double-face observation, taken with a 1″ theodolite, is in the region of ±2.5″.


## 4.4.3 Whole-circle bearings

The next step in the computational procedure is that of finding the whole-circle bearings (WCB) from the *corrected* angles.

WCB are illustrated in *Figure 4.17(a)*. The angles are always measured clockwise from the zero axis and range from 0° to 360°, thus:

WCB of *PA* =   40°
WCB of *PB* = 120°
WCB of *PC* = 250°
WCB of *PD* = 330°

Purely as a matter of historical interest, the above WCB are here expressed as equivalent *quadrant bearings* (QB) in *Figure 4.17(b)*:

QB of *PA* = N 40° E
QB of *PB* = S  60° E
QB of *PC* = S  70° W
QB of *PD* = N 30° W

The universal use of electronic calculators has now rendered the quadrantal-bearing system obsolete.

The WCB of the traverse lines are obtained by adding the measured angles to the previous WCB. The starting leg of a traverse is generally given a local arbitrary WCB of 0° 00′ 00″, unless it commences from existing surveys. Students should become familiar with the reduction of angles to bearings and *vice versa*, from first principles. The following approach is suggested.

*Example 1.* Find the WCB of *PB*, given:

WCB of *AP* =   0° 00′ 00″,   clockwise angle *APB* = 120° 00′ 00″
WCB of *AP* = 89° 35′ 36″,   clockwise angle *APB* = 104° 10′ 10″

WCB of $AP$ = 348° 20′ 20″,　clockwise angle $APB$ = 300° 00′ 00″
WCB of $AP$ =　08° 10′ 10″,　clockwise angle $APB$ = 285° 50′ 40″

## Method

(1) Always work from the point about which the angle is measured, namely $P$. If WCB of $AP$ is 0°, then the reverse bearing $PA$ is 180°; now simply add the clockwise angle $APB$ to give the WCB of $PB$, i.e. 180° + 120° = 300°. At first a sketch is very useful (see *Figure 4.18(a)*).

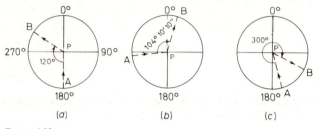

Figure 4.18

(2) *Figure 4.18(b)*: Reverse bearing $AP$ = (89° 35′ 36″ + 180°) = 269° 35′ 36″. ∴ WCB of $PB$ = (269° 35′ 36″ + 104° 10′ 10″) = 373° 45′ 46″ = 13° 45′ 46″.
(3) *Figure 4.18(c)*: Reverse bearing $AP$ = (348° 20′ 20″ − 180°) = 168° 20′ 20″. ∴ WCB of $PB$ = (168° 20′ 20″ + 300° 00′ 00″) = 468° 20′ 20″ = 108° 20′ 20″.

*Note:* In the above two instances $PB$ swings through 360° to give, say, 373°; as the bearing cannot be greater than 360°, it has swung to position (373° − 360°) = 13°.

(4) The student should attempt this for himself, the result being WCB of $PB$ = 114° 00′ 50″.

In the same way, students should be able to calculate angles when given the WCB of two lines. Using WCB, the method is simply the reverse of the previous approach.

*Example 2.* Find the clockwise angle, given:

WCB of $AP$ =　　0° 00′ 00″,　of $PB$ = 300° 00′ 00″
WCB of $AP$ =　89° 35′ 36″,　of $PB$ =　13° 45′ 46″
WCB of $AP$ = 348° 20′ 20″,　of $PB$ = 108° 20′ 20″
WCB of $AP$ =　08° 10′ 10″,　of $PB$ = 114° 00′ 50″

## Method

Again work from the angle point $P$; thus if $AP$ = 0° then $PA$ = 180° and $PB$ = 300°. ∴ angle $APB$ = (300° − 180°) = 120°.

Students should now work the rest for themselves, obtaining the answers as given in the previous questions.

It will now be apparent that to carry out the above for a traverse of many stations would be extremely tedious, and the following method is used. From *Figure 4.19* the following information may be deduced:

WCB of $AB = \theta_A$
Measured angle $ABC = \alpha$
WCB of $BC = \theta_B$
$\therefore \ \theta_B = \theta_A + \alpha - 180°$   i.e.

WCB of $BC$ = WCB of $AB$ + measured angle $- 180°$ giving the following rules which the student should memorize:

If the sum of the previous WCB and the measured angle is *greater* than 180°, then *subtract* 180°: if *less* than 180° *add* 180°. If the sum is *greater than 540° subtract 540°*.

*Example 3.* The internal clockwise angles of a closed polygonal traverse are as shown in *Table 4.1.* Correct them and tabulate the bearings, given WCB of $AB = 0°\ 00'\ 00''$.

The WCB are deduced as follows, using the given rule:

| | |
|---|---|
| WCB of $AB$ = | 0° 00′ 00″ |
| Angle $ABC$ = | 120° 20′ 05″ |
| | 120° 20′ 05″ |
| | +180° |
| WCB of $BC$ = | 300° 20′ 05″ |
| Angle $BCD$ = | 86° 00′ 45″ |
| | 386° 20′ 50″ |
| | −180° |
| WCB of $CD$ = | 206° 20′ 50″ |
| Angle $CDE$ = | 341° 34′ 25″ |
| | 547° 55′ 15″ |
| | −540° |
| WCB of $DE$ = | 07° 55′ 15″ |
| Angle $DEF$ = | 60° 22′ 05″ |
| | 68° 17′ 20″ |
| | +180° |
| WCB of $EF$ = | 248° 17′ 20″ |
| Angle $EFA$ = | 100° 28′ 25″ |
| | 348° 45′ 45″ |
| | −180° |
| WCB of $FA$ = | 168° 45′ 45″ |
| Angle $FAB$ = | 11° 14′ 15″ |
| | 180° 00′ 00″ |
| | −180° |
| WCB of $AB$ | 0° 0′ 0″   (Check) |

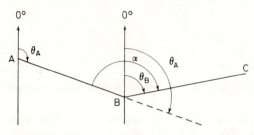

*Figure 4.19*

## 4.4.4 Plane rectangular co-ordinates

Using the WCB and the horizontal lengths of the traverse legs, the plane rectangular co-ordinates of each traverse station are now computed. The co-ordinate values are then used

(1) In the adjustment of the traverse.
(2) To plot accurately the positions of the stations on a precisely-constructed grid.
(3) To carry out setting-out computations.

For surveys of limited extent, *plane rectangular co-ordinates* are used. With reference to *Figure 4.20* the positions of the traverse stations *A*, *B*, *C* and *D* are fixed by perpendicular measurements from a rectangular axis.

Distances along the horizontal axis (X-axis) are termed *eastings* (*E*). Distances along the vertical axis (in pure maths the Y-axis) are termed *northings* (*N*). The usual mathematical sign convention is used, that is, distances to the north and east of the origin are *positive*; to the south and west *negative*. Students used to the normal mathematical convention of defining polar co-ordinates by the angle measured from the +E axis, must remember that the WCB of a line is measured from the +N axis.

On *Figure 4.20*, the *difference in co-ordinates* of *B* relative to *A* are obtained from the right-angled triangle *AaB*, i.e.

$$aB = \Delta E = L \sin \alpha \tag{4.4}$$

$$Aa = \Delta N = L \cos \alpha \tag{4.5}$$

*Figure 4.20*

where $L$ is the horizontal distance and $\alpha$ the WCB of the line. Algebraic addition of the co-ordinate differences produces the *total co-ordinates, E and N*, of the points relative to the origin. The above operation using equations (4.4) and (4.5) is known as *computing the polar*.

From *Figure 4.20* it can be seen that the co-ordinates of $D$ relative to $A$ are $+263$ $E$ and $-25$ $N$, from which it is possible to *compute the join*, i.e. the length and bearing of $AD$.

Since the computation of the join and polar are fundamental to surveying, they will now be illustrated in detail. To facilitate understanding of the procedures, students should study *Figure 4.21(a)* and *Table 4.2*, noting that the signs of the co-ordinates define the quadrant in which the WCB lies, and *vice versa*.

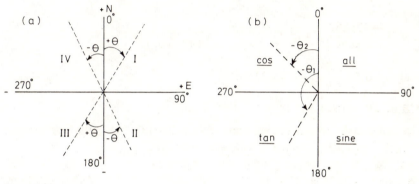

Figure 4.21

TABLE 4.2

| Bearing | E | N |
|---|---|---|
| Quadrant I | + | + |
| Quadrant II | + | − |
| Quadrant III | − | − |
| Quadrant IV | − | + |

(1) The *join* is the length ($L$) and bearing of a line obtained from the difference in co-ordinates of its ends.

Consider two points $A$ and $B$ whose co-ordinates are $E_A$, $N_A$ and $E_B$, $N_B$, then:

$$\Delta E_{AB} = (E_B - E_A) \quad \text{and} \quad \Delta N_{AB} = (N_B - N_A)$$

and, from the basic equations (4.4) and (4.5):

$$\alpha_{AB} = \tan^{-1}\frac{\Delta E}{\Delta N} = \cot^{-1}\frac{\Delta N}{\Delta E} \qquad (4.6)$$

$$L_{AB} = (\Delta E^2 + \Delta N^2)^{1/2} = \frac{\Delta E}{\sin \alpha} = \frac{\Delta N}{\cos \alpha} \qquad (4.7)$$

In the following computations, it is assumed that a scientific pocket calculator is used.

## Example

$$E_A = 48964.38 \text{ m} \qquad N_A = 69866.75 \text{ m}$$
$$E_B = 48988.66 \text{ m} \qquad N_B = 62583.18 \text{ m}$$

---

$$\Delta E_{AB} = \; +24.28 \text{ m} \qquad \Delta N_{AB} = -7283.57 \text{ m}$$

---

$$\alpha_{AB} = \tan^{-1} \frac{+24.28}{-7283.57} = -0° \, 11' \, 27''$$

From the signs of $\Delta E$, $\Delta N$ ($+ \, -$), the bearing lies in quadrant II

$$\therefore \; \text{WCB of } \overline{AB} = 180° - 0° \, 11' \, 27'' = 179° \, 48' \, 33''$$

*Check:*

$$\alpha_{AB} = \cot^{-1} \frac{-7283.57}{+24.28} = -0° \, 11' \, 27''$$

Alternatively, $\quad \cot^{-1} \dfrac{\Delta E}{\Delta N} = 89° \, 48' \, 33''$

$$\therefore \; \text{WCB of } AB = 90° + 89° \, 48' \, 33'' = 179° \, 48' \, 33''$$

In this case, it is easier to add $\alpha$ to 90° than subtract from 180° as in the first instance.

$$L_{AB} = (\Delta E^2 + \Delta N^2)^{1/2} = (24.28^2 + 7283.57^2)^{1/2} = 7283.61 \text{ m}$$
$$= \Delta N/\cos \alpha = 7283.57/\cos 179° \, 48' \, 33'' = 7283.61 \text{ m}$$
$$= \Delta E/\sin \alpha = 24.28/\sin 179° \, 48' \, 33'' = 7289.84 \text{ m} \qquad \text{(error} = 6.23 \text{ m)}$$

The sines and tangents of small angles ($<1° \, 20'$) and the cosines of large angles ($>88° \, 40'$) vary greatly and erratically, thus any rounding-off error will have a greater effect on the sine value above than on the cosine. This then is the reason for the large error in distance when using $\Delta E/\sin \alpha$. Thus the use of Pythagoras is recommended with calculators, or, when using either of the latter equations, choose the one which has the largest co-ordinate difference, i.e. $\Delta N > \Delta E$; therefore, in this instance, use $\Delta N/\cos \alpha$.

(2) The *polar* is the co-ordinates of point $B$ given the co-ordinates of $A$ and the length and bearing of $AB$, thus:

$$E_B = E_A + \Delta E_{AB}$$
$$N_B = N_A + \Delta N_{AB}$$

where $\quad \Delta E = L \sin \alpha$
$\qquad\quad\; \Delta N = L \cos \alpha$

e.g. $\; E_A = 48 \, 964.38 \text{ m}, \qquad N_A = 69 \, 866.75 \text{ m}$

$\quad$ WCB of $A - B = 299° \, 58' \, 46''$
$$L_{AB} = 1325.64 \text{ m}$$

As $AB$ is in quadrant IV, then the signs of $\Delta E$ and $\Delta N$ are $-$ and $+$ respectively

$$\therefore \; \Delta E_{AB} = 1325.64 \sin 299° \, 58' \, 46'' = -1148.28 \text{ m}$$
$$\Delta N_{AB} = 1325.64 \cos 299° \, 58' \, 46'' = +662.41 \text{ m}$$
$$\therefore \; E_B = E_A + \Delta E_{AB} = 47 \, 816.10 \text{ m}$$
$$N_B = N_A + \Delta N_{AB} = 70 \, 529.16 \text{ m}$$

As pocket calculators obey that well-known law of trigonometry illustrated by *Figure 4.21(b)*, the correct signs of the $\Delta E$, $\Delta N$ values will be automatically supplied in the readout. Whilst correctly-working and fully-charged pocket calculators do not make computational mistakes there is still a need to check the input operation. To this end the 'auxiliary bearings' check can be used, as follows:

$$\sqrt{2} \times L \sin(\alpha + 45°) = \Delta N + \Delta E$$
$$\sqrt{2} \times L \cos(\alpha + 45°) = \Delta N - \Delta E$$

$$\therefore \quad \sqrt{2} \times 1325.64 \sin 344° \, 58' \, 46'' = -485.87 = \Delta N + \Delta E$$
$$\sqrt{2} \times 1325.64 \cos 344° \, 58' \, 46'' = +1810.69 = \Delta N - \Delta E$$

| | | | | |
|---|---|---|---|---|
| Adding: | $+2 \times \Delta N = +1324.82$ | | $\therefore \quad \Delta N = +662.41$ m | |
| Subtract: | $+2 \times \Delta E = -2296.56$ | | $\therefore \quad \Delta E = -1148.28$ m | |

It is worth noting that if the calculator possesses the *polar* and *rectangular* co-ordinate keys, generally denoted by $\rightarrow P$ and $\rightarrow R$, then inputting rectangular co-ordinates of the III and IV quadrants and changing to polars, will give as the angular value $-\theta_1$ and $-\theta_2$ as shown in *Figure 4.21(b)*. Thus, to obtain the WCB one must add 360°. The reverse operation is not affected.

However, if the normal trig function ($\tan^{-1} \Delta E/\Delta N$) is used, the angular values will be given as in *Figure 4.21(a)* and must be treated accordingly, in quadrants II, III and IV, to convert them to WCB.

Similarly, in computer programming, the use of $ATN$ ($\Delta E/\Delta N$) produces the situation as in *Figure 4.21(a)*. One should therefore use ATN2($\Delta E, \Delta N$) which produces the much easier situation as in *Figure 4.21(b)*.

As a guide to the number of decimal places to use when computing, the following is usual:

$$21° \, 21' \, 00'' \quad = 4 \text{ dec places}$$
$$21° \, 21' \, 10'' \quad = 5 \text{ dec places}$$
$$21° \, 21' \, 11'' \quad = 6 \text{ dec places}$$
$$21° \, 21' \, 11.1'' = 7 \text{ dec places}$$

## 4.4.5 Traverse adjustment

The final step in the computational procedure is the adjustment of the *co-ordinate differences* ($\Delta E$, $\Delta N$) in order to make the network geometrically correct.

In the case of the polygon the algebraic sum of the $\Delta E$ and $\Delta N$ values should, theoretically, equal zero, if not, the discrepancies ($\Delta'E$, $\Delta'N$) are distributed throughout the network.

In the case of the link traverse, the algebraic sum of the $\Delta E$ and $\Delta N$ values relative to the origin, should, theoretically, equal the co-ordinate values of the station into which the traverse is connected.

The methods of adjustment in popular use vary from the very elementary to the very advanced. However, regardless of the mathematical rigour used, there is no ideal method of adjustment. Of the elementary methods, the most popular is the *Bowditch method* which states:

$$(1) \quad \delta E_i = \frac{\Delta'E}{\displaystyle\sum_{i=1}^{n} L_i} \cdot L_i = K_1 \cdot L_i \qquad (2) \quad \delta N_i = \frac{\Delta'N}{\displaystyle\sum_{i=1}^{n} L_i} \cdot L_i = K_2 \cdot L_i$$

where,   $\delta E_i, \delta N_i$ = the co-ordinate corrections

$\Delta'E, \Delta'N$ = the co-ordinate misclosure (constant)

$L_i$ = the horizontal length of the $i$th traverse leg

$\sum_{i=1}^{n} L_i$ = the sum of the lengths (constant)

$K_1$ & $K_2$ = the resultant constants

This method, although devised by Nathaniel Bowditch as long ago as 1807 for the adjustment of compass traverses, nevertheless remains the most popular method in use today.

The application of the method to a polygonal traverse (*Figure 4.22*) will now be demonstrated.

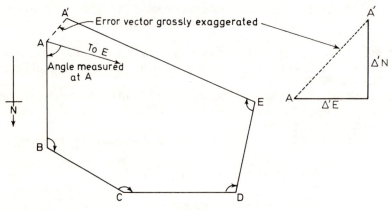

*Figure 4.22*

**TABLE 4.3**

| Line | $\Delta E$ | $\Delta N$ |
|------|-----------|-----------|
| AB   | —         | +155.00   |
| BC   | −172.44   | +101.31   |
| CD   | −249.00   | —         |
| DE   | −18.87    | −189.06   |
| EA   | +439.76   | −68.08    |
| Sum  | −0.55     | −0.83     |

The measured field data are shown in *Table 4.5* where the angles have been corrected and reduced to WCB. Using the WCB and horizontal lengths the co-ordinate differences are computed using equations (4.4) and (4.5) and they are shown in *Table 4.3*.

Algebraic summation of the co-ordinates shows the errors

$$\Delta'E = -0.55; \qquad \Delta'N = -0.83$$

which are distributed throughout the traverse using the Bowditch rule, as shown in

**TABLE 4.4**

| Corrections to $\Delta E$ | Correction to $\Delta N$ |
|---|---|
| $AB = \dfrac{+0.55}{1239} \times 155$ | $= \dfrac{+0.83}{1239} \times 155$ |
| $= K_1 \times 155 = +0.07$ | $= K_2 \times 155 = +0.10$ |
| $BC = K_1 \times 200 = +0.09$ | $= K_2 \times 200 = +0.13$ |
| $CD = K_1 \times 249 = +0.11$ | $= K_2 \times 249 = +0.17$ |
| $DE = K_1 \times 190 = +0.08$ | $= K_2 \times 190 = +0.13$ |
| $EA = K_1 \times 445 = +0.20$ | $= K_2 \times 445 = +0.30$ |
| Sum $= +0.55$ | Sum $= +0.83$ |

*Table 4.4*. Note that as the co-ordinate misclosure is *negative*, the corrections are *positive*.

The above corrections are added algebraically to the co-ordinate differences to produce the corrected values shown within the thick blocks of *Table 4.5*.

Finally, algebraic summation of the corrected $\Delta E$ and $\Delta N$ values gives the *total co-ordinates* ($E$ and $N$) of the survey stations $B$, $C$, $D$ and $E$ relative to the origin $A$.

The total amount of misclosure, $AA'$, is termed the *error vector*. Its length and bearing are calculated using equations (4.7) and (4.6), and it is shown in *Table 4.5*. Expressing the error vector as a proportion of the total length of the traverse is the manner of expressing the accuracy of the traverse, i.e. 1 in 1239. Engineering surveys may range in accuracy from 1 in 5000 to 1 in 50 000, depending upon the project specifications.

The total co-ordinates may now be plotted on an accurately-constructed grid and used as a reference framework from which the topographic detail is plotted.

An examination of the *Worked examples* will show the application of co-ordinate computations in typical engineering situations.

## 4.4.6 Computational accuracy

In carrying out computation, it should be realized that computation cannot improve the accuracy of the captured field data.

For instance, in the previous examples, angles were measured to the nearest 10″ and distances to the nearest 0.01 m; thus the co-ordinate differences should not be computed to any greater accuracy than two places of decimals. With pocket calculators, accuracy may be to eight or nine places of decimals, in which case the answer must be rounded down to two places. The final computations can only be as accurate as the least accurate data used.

Also, when rounding-off numbers a set procedure should be adhered to. For example, 103.5 may be rounded up to 104, whilst 102.5 is rounded down to 102. This results in the round-off error being randomized. If one always rounds up from 0.5, the error becomes systematic and accumulative. Thus the rule is: If the number to be rounded-off is *odd*—round *up*, if *even*—round *down*.

**TABLE 4.5. A Bowditch adjustment of closed polygonal traverse**

| Angle | Observed horizontal angle (° ′ ″) | Corrn | Corrected horizontal angle (° ′ ″) | Line | WCB (° ′ ″) | Horizontal length (m) | ΔE + | ΔE − | ±Corr | Corrected | ΔN + | ΔN − | ±Corr | Corrected | Total co-ord E | Total co-ord N | Stns |
|---|---|---|---|---|---|---|---|---|---|---|---|---|---|---|---|---|---|
| | | | | | | | | | | | | | | | 0.0 | 0.0 | A |
| AB̂C | 120 25 50 | +10″ | 120 26 00 | AB | 0 00 00 (assumed) | 155 | | — | +0.07 | 0.07 | 155.00 | — | +0.10 | 155.10 | +0.07 | +155.10 | B |
| BĈD | 149 33 50 | +10″ | 149 34 00 | BC | 300 26 00 | 200 | | 172.44 | +0.09 | 172.35 | 101.31 | | +0.13 | 101.44 | −172.28 | +256.54 | C |
| CD̂E | 95 41 50 | +10″ | 95 42 00 | CD | 270 00 00 | 249 | | 249.00 | +0.11 | 248.89 | — | | +0.17 | 0.17 | −421.17 | +256.71 | D |
| DÊA | 93 05 50 | +10″ | 93 06 00 | DE | 185 42 00 | 190 | | 18.87 | +0.08 | 18.79 | | 189.06 | +0.13 | 188.93 | −439.96 | +67.78 | E |
| EÂB | 81 11 50 | +10″ | 81 12 00 | EA | 98 48 00 | 445 | 439.76 | | +0.20 | 439.96 | | 68.08 | +0.30 | 67.78 | 0.00 | 0.00 | A |
| Sum | 539 59 10 | +50″ | 540 00 00 | AB | 0 00 00 | Sum = 1239 | +439.76 | −440.31 | | | +256.31 | −257.14 | | | Check | | |
| (2n − 4)90° | 540 00 00 | | Check | | Check | | Error | | −0.55 | | −0.83 | | | | | | |
| Error | | −50 | | | | | Corr | | +0.55 | | +0.83 | | | | | | |

Error vector = $(0.55^2 + 0.83^2)^{1/2}$ = 1 m (213° 32′)
Proportional error = 1 in 1239

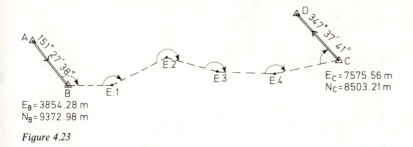

Figure 4.23

### 4.4.7 Link traverse adjustment

A link traverse (*Figure 4.23*) commences from known stations (*AB*) and connects into known stations (*CD*). Stations *A*, *B*, *C*, *D*, are usually fixed to a higher order of accuracy, with values which remain unaltered in the subsequent computation.

The method of computation and adjustments proceeds as follows:

#### Angular adjustment

(1) Compute the WCB of *CD* through the traverse from *AB* and compare it with the known bearing of *CD*. The difference ($\Delta$) of the two bearings is the angular misclosure.
(2) As a *check* on the value of $\Delta$ the following rule may be applied. Computed WCB of *CD* = (Sum of observed angles + initial bearing (*AB*)) $- n \cdot 180°$ where *n* is the number of angles and is $+$even, $-$odd. If the result is negative, add $360°$.
(3) The correction per angle would be $\Delta/n$, which is distributed *accumulatively* over the WCB as shown in the example.

#### Co-ordinate adjustment

(1) Compute the total co-ordinates of *C* through the traverse from *B* as origin. Comparison with the known co-ordinates of *C* gives the co-ordinate misclosure $\Delta'E$, and $\Delta'N$.
(2) As the computed co-ordinates are *total* values, distribute the misclosure *accumulatively* over stations *E*.1 to *C*.

Now study the example given in *Table 4.6*.

## 4.5 SOURCES OF ERROR

At this stage the reader should have a clearer understanding of the field data required for the fixing of control stations by the measurement of horizontal and vertical angles and distances; i.e., theodolite traversing.

To obtain the angles the instrument and targets must be carefully centred over their respective survey stations.

The distances may be obtained by electromagnetic processes, by optical processes, or by direct taping with a steel band. Since the electromagnetic processes are dealt with in

**TABLE 4.6 Bowditch adjustment of a link traverse**

| Stns | Observed angles (° ' ") | Line | WCB (° ' ") | Corrn (") | Adjusted WCB (° ' ") | Dist (m) | Unadjusted E | Unadjusted N | Corrn δE | Corrn δN | Adjusted E | Adjusted N | Stn |
|---|---|---|---|---|---|---|---|---|---|---|---|---|---|
| A | | A–B | 151 27 38 | | 151 27 38 | | 3854.28 | 9372.98 | | | 3854.28 | 9372.98 | B |
| B | 143 54 47 | B–E1 | 115 22 25 | −4 | 115 22 21 | 651.16 | 4442.63 | 9093.96 | +0.03 | −0.05 | 4442.66 | 9093.91 | E.1 |
| E.1 | 149 08 11 | E1–E2 | 84 30 36 | −8 | 84 30 28 | 870.92 | 5309.55 | 9177.31 | +0.08 | −0.11 | 5309.63 | 9177.20 | E.2 |
| E.2 | 224 07 32 | E2–E3 | 128 38 08 | −12 | 128 37 56 | 522.08 | 5171.38 | 8851.36 | +0.11 | −0.15 | 5171.49 | 8851.21 | E.3 |
| E.3 | 157 21 53 | E3–E4 | 106 00 01 | −16 | 105 59 45 | 1107.36 | 6781.87 | 8546.23 | +0.17 | −0.22 | 6782.04 | 8546.01 | E.4 |
| E.4 | 167 05 15 | E4–C | 93 05 16 | −20 | 93 04 56 | 794.35 | 7575.35 | 8503.49 | +0.21 | −0.28 | 7575.56 | 8503.21 | C |
| C | 74 32 48 | C–D | 347 38 04 | −23 | 347 37 41 | | 7575.56 | 8503.21 | | | | | |
| | | C–D | 347 37 41 | +23 | Sum = | 3946.15 | | | | | | | |
| D | | Δ | | | | | | | | | | | |

Δ'E, Δ'N = −0.21  +0.28

Unadjusted: 7575.56  8503.21

Bottom-left summary:

| | ° ' " |
|---|---|
| Sum | 916 10 26 |
| Initial bearing | 151 27 38 |
| Total | 1067 38 04 |
| −6 × 180° | 1080 00 00 |
| | −12 21 56 |
| | +360 00 00 |
| CD (comp) | 347 38 04 |
| CD (known) | 347 37 41 |
| Δ | +23  Check |

Right-hand computations:

Error vector $= (0.21^2 + 0.28^2)^{1/2} = 0.35$

Proportional error $= \dfrac{0.35}{3946} = 1/11\,300$

Volume 2, and the optical processes are dealt with in the next Chapter, only the errors of taping will be dealt with here.

The sources of error in traversing, therefore, are:

(1) Angular errors.
(2) Centring errors.
(3) Linear errors.

### 4.5.1 Angular errors

Instrumental errors have already been dealt with in detail. It was shown that, with the exception of plate bubble error, double-face observations would eliminate the majority of these error sources.

(1) *Sighting:* due to imperfections of human sight and touch, target bisections are rarely perfect. These errors are just as likely to be positive as negative. Their effect is reduced by taking the mean of several measurements.
(2) *Reading and setting verniers:* this again is a human error, and can be reduced by taking the mean of several readings.
(3) *Instrument operation* such as turning the wrong tangent screw, failure to eliminate parallax, leaving lower plate unclamped, tripod movement. All these errors should become apparent when booking the observations.
(4) *Booking error* can be eliminated by the booker reading the observation back to the observer.
(5) *Natural causes* such as the effect of shimmer, refraction, wind, and differential expansion of the instrument's parts. Little can be done in the first two instances, but in the latter two, the instrument·tripod should be very firmly established and shielded from wind and sun.

### 4.5.2 Centring errors

Errors in the centring of the theodolite and targets directly affect the measured horizontal angle.

The effect of defective centring of the theodolite is clearly illustrated in *Figure 4.24.* If the theodolite was centred at $B'$ instead of $B$, the angle $AB'C$ would be measured instead

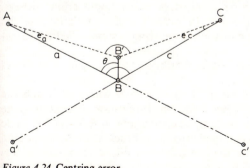

*Figure 4.24* Centring error

of the required angle $ABC$. The error would be equal to the sum of the angles subtended by $BB'$ (the centring error) at $A$ and $C$ and is positive in sign, i.e. $+(e_a + e_c)$. If $B'$ falls in sector $a'Bc'$, the error is $-(e_a + e_c)$. If $B'$ falls within sectors $aBa'$ and $cBc'$, the error is the *difference* of the subtended angles and is positive or negative depending on which side of an arc through $ABC$ it falls.

In a similar way, the defective centring of the targets at $A$ and $C$ can be considered, although their combined effect is only half that of the theodolite error.

If $AB = 200$ m, then a centring error of 1 mm $(BB')$ would subtend an angle of $1''(e_a)$ at $A$. However, if $AB$ was 20 m then $e_a = 10''$, and assuming $AB = BC$ and so $e_a = e_c$, the angular error would be $20''$. This shows that the effect increases as the length of the traverse leg decreases.

The combined effect of target and theodolite centring errors are shown in *Figure 4.25* for centring errors of $\sigma = \pm 4$ mm, $\pm 2$ mm and $\pm 1$ mm and various combinations of traverse leg lengths $(L_1, L_2)$ subtending the angle.

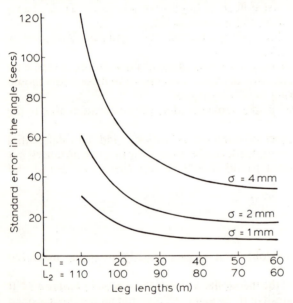

Figure 4.25

The inclusion of short lines cannot be avoided in many engineering surveys, particularly in underground tunnelling work. In order therefore to minimize the propagational effect of centring error, a constrained centring system called the *three-tripod system* (TTS) is used.

The TTS uses interchangeable levelling heads or tribrachs and targets, and works much more efficiently with a fourth tripod.

Consider *Figure 4.26*; tripods are set up at $A$, $B$, $C$ and $D$ with the detachable tribrachs carefully levelled and centred over each station. Targets are clamped into the tribrachs at $A$ and $C$, whilst the theodolite is clamped into the one at $B$. When the angle $ABC$ has been measured, the target $A$ is clamped into the tribrach at $B$, the theodolite into the tribrach at $C$ and that target into the tribrach at $D$. Whilst the angle $BCD$ is being measured, the tripod and tribrach is removed from $A$ and set up at $E$ in

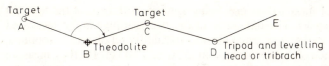

*Figure 4.26*

preparation for the next move forward. This technique not only produces maximum speed and efficiency, but also confines the centring error to the station at which it occurred. Indeed, the error in question here is not one of centring in the conventional sense, but one of knowing whether or not the central axis of the targets and theodolite, when moved forward, occupy exactly the same positions as did their previous occupants.

The confining of centring errors using the above system can be explained by reference to *Figure 4.27*. Consider first the use of the TTS. The target erected at $C$, 100 m from $B$, is badly centred resulting in a displacement of 50 mm to $C'$. The angle measured at $B$ would be $ABC'$ in error by $e$. The error $e$ is 1 in 2000 $\approx$ 2 min (*N.B.* If $BC$ was 10 m long then $e = 20$ min.)

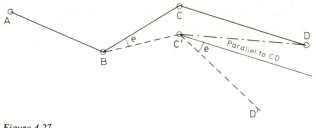

*Figure 4.27*

The target is now removed from $C'$ and replaced by the theodolite which measures angle $BC'D$, thus bringing the survey back onto $D$. The only error would therefore be a co-ordinate error at $C$ equal to the centring error and would obviously be much less than the grossly exaggerated 50 mm used here.

Consider now conventional equipment using one tripod and theodolite and sighting to ranging rods. Assume the rod at $C$ due to bad centring or tilting appears to be at $C'$, the wrong angle $ABC'$ would be measured. Now, when the theodolite is moved it would this time be correctly centred over the station at $C$ and the correct angle $BCD$ measured. However, this correct angle would be added to the previous computed bearing, which would be that of $BC'$, giving the bearing $C'D'$. Thus the error $e$ is propagated from the already incorrect position at $C'$ producing a further error at $D'$ of the traverse. Centring of the instrument and targets precisely over the survey stations is thus of paramount importance.

## 4.5.3 Linear errors

The first step in the taping process when measuring a line $AB$ is to fix measuring marks exactly in line with $A$ and $B$ at intervals less than a tape length. These intervals are sometimes referred to as *bays*.

To measure a bay, the tape is laid between the two marks, carefully aligned and tensioned. At the instant of standard tension being applied, both ends of the tape are read against the pre-set marks. The difference of the two readings is the measured length. The process is repeated several times until a satisfactory set of data is obtained.

Most steel tapes and bands are *standardized* at 20°C and 10 kgf; i.e. they measure their prescribed length under these conditions. Thus for very high accuracy the tape temperature will need to be measured and a correction applied for temperature variations above or below standard.

The error sources are as follows:

## (1)  Standardization

Tapes may be stretched due to continual over-tensioning or shortened due to kinks or defective repairs. If too long or too short by an amount $\delta L$, then a systematic error equal to this amount will occur each time the tape is laid down.

The amount of this error can be ascertained by comparing the working tape with a new tape kept especially for this purpose. Some government agencies will also carry out this standardization procedure, using very sophisticated procedures such as laser interferometry, on steel or invar bands.

For example, a tape found on standardization to be 30.01 m in length will still record 30 m when laid down, but the recorded measurement of 30 m will require a positive correction of 0.01 m. Therefore, the rule is that when the tape is too *long* the correction is *positive*, and *vice versa*.

## (2)  Temperature

If the tape temperature is recorded during measurement and found to differ from the standard (20°C) a correction may be computed from

$$C_T = L.K.\Delta T$$

where   $L$ = measured length (m)
$\quad\quad\quad K$ = co-efficient of expansion (steel = $11.2 \times 10^{-6}$ per °C)
$\quad\quad\quad \Delta T$ = temperature difference from standard (°C)

## (3)  Tension

If the correct tension is applied, with either a tension handle or a spring balance, no correction is necessary. It may sometimes be necessary to over-tension, however, and the correction may be calculated using:

$$C_t = L.\Delta t/A.E$$

where   $\Delta t$ = difference in tension from standard
$\quad\quad\quad A$ = cross-sectional area of tape
$\quad\quad\quad E$ = Young's modulus of elasticity

In the above formula, the units must be compatible; thus, for $C_t$ in metres, $L_2$ must be in metres (m), $\Delta t$ in newtons (N), $A$ in mm$^2$ and $E$ in N/mm$^2$.

## (4) Sag

If the tape is not supported throughout its length it will sag. If the sag cannot be eliminated by field procedure, i.e. building support, etc., a correction may be computed from:

$$C_s = \omega^2 L^3 / 24 t^2$$

where    $\omega$ = mass per unit length
$t$ = *applied* tension

The units must be compatible.

## (5) Slope

If the distance is measured on the slope it may be reduced to the horizontal using Pythagoras, if the difference between the heights of the ends is known; or by using $L(1 - \cos \theta)$ if the vertical angle $\theta$ is known.

## (6) Faulty alignment

This results in too long a measurement with a systematic error of $d^2/2L$, where $d$ is the displacement per length $L$. Careful alignment by eye is generally sufficient.

## (7) Surface irregularities

This results in a vertical deformation producing a systematic error of $2h^2/L$, where $h$ is the maximum deformation at the centre. The tape may be laid along planks of wood, carefully placed in position, or over-tensioned, to reduce this error.

## (8) Errors in reading and marking the tape

These are of a random nature and are minimized by taking the mean of several measurements.

## (9) Errors in booking data

Also of a random nature, but easily recognized when several measures of the same length are obtained. (For a more detailed discussion on taping, refer to Volume 2.)

## 4.6 LOCATION OF GROSS ERROR

In the case of a single gross error or mistake in either angle or distance, its position may be located and re-measured in the field.

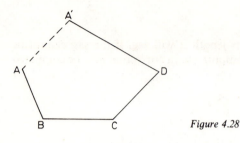

*Figure 4.28*

## 4.6.1 Gross error in distance

Occurs in the bay having approximately the same bearing as the error vector. For instance, in *Figure 4.28* the gross error resulting in the error vector $AA'$ has obviously occurred in line $CD$. Thus reducing line $CD$ by the amount $AA'$ would cause $A'$ to approach $A$.

## 4.6.2 Gross error in angle

Can be detected by calculating the co-ordinate values of the stations twice; once by commencing from the known bearing $AB$, as in *Figure 4.23*, say, and then using the bearing $CD$ and working back to $A$. The error will most probably have occurred at the station where the co-ordinate values approximately agree. The unadjusted angles are used in the computation. Alternatively, the traverse may be plotted in both directions to locate the necessary station.

Where there is more than one error these techniques will not function. Students should not confuse these gross errors with the normal accidental errors which are distributed by adjustment.

## 4.7 AREAS BY CO-ORDINATES

The area enclosed by the traverse $ABCDA$ in *Figure 4.20* can be found by taking the area of the rectangle $a'cDd$ and subtracting the surrounding triangles, etc.,as follows:

$$
\begin{aligned}
\text{Area of rectangle } a'cDd = a'c \times a'd & \\
= 263 \times 173 &= 45\,499 \quad \text{m}^2 \\
\text{Area of rectangle } a'bBa = \quad 77 \times 71 \quad &= \quad 5\,467 \quad \text{m}^2 \\
\text{Area of triangle } AaB \quad = \quad 71 \times 35.5 &= \quad 2\,520.5 \; \text{m}^2 \\
\text{Area of triangle } BbC \quad = \quad 77 \times 46 \quad &= \quad 3\,542 \quad \text{m}^2 \\
\text{Area of triangle } CcD \quad = 173 \times 50 \quad &= \quad 8\,650 \quad \text{m}^2 \\
\text{Area of triangle } DdA \quad = 263 \times 12.5 &= \quad 3\,287.5 \; \text{m}^2 \\
\end{aligned}
$$

$$\text{Total} = 23\,467 \quad \text{m}^2$$

$\therefore$ Area $ABCDA = 45\,499 - 23\,467 = 22\,032$ m$^2 \approx 22\,000$ m$^2$

The following rule may be used when the *total co-ordinates* only are given. Multiply the algebraic *sum* of the northing of each station and the one following by the algebraic *difference* of the easting of each station and the one following. The area is half the

TABLE 4.7

| Stns | E | N | Difference of E | Sum of N | Double area + | Double area − |
|------|------|------|------|------|------|------|
| A | 0.0 | 0.0 | −71 | 71 | | 5 041 |
| B | 71 | 71 | −92 | 219 | | 20 148 |
| C | 163 | 148 | −100 | 123 | | 12 300 |
| D | 263 | −25 | 263 | −25 | | 6 575 |
| A | 0.0 | 0.0 | | | | — |

$$\Sigma \qquad 44\,064$$

Area $ABCDA = 22\,032\ \text{m}^2 \approx 22\,000\ \text{m}^2$

algebraic sum of the products. Thus, from *Table 4.7, Figure 4.20*

Area  $ABCDA = 22\,032\ \text{m}^2 \approx 22\,000\ \text{m}^2$

The value of 22 000 m² is more correct considering the number of significant figures involved in the computations.

This latter rule is the one most commonly used and is easily remembered if written as follows:

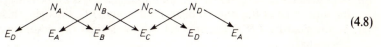

(4.8)

Thus  $A = 0.5[N_A(E_B − E_D) + N_B(E_C − E_A) + N_C(E_D − E_B)$
$+ N_D(E_A − E_C)]$
$= 0.5[0 + 71(163) + 148(263 − 71) + −25(0 − 163)]$
$= 0.5[11\,573 + 28\,416 + 4075] = 22\,032\ \text{m}^2$

## 4.8 PARTITION OF LAND

This task may be carried out by an engineer when sub-dividing land either for large building plots or for sale purposes.

### 4.8.1 To cut off a required area by a line through a given point

With reference to *Figure 4.29*, it is required to find the length and bearing of the line *GH* which divides the area *ABCDEFA* into the given values.

*Method*

(1) Calculate the total area *ABCDEFA*.
(2) Given point *G*, draw a line *GH* dividing the area approximately into the required portions.
(3) Draw a line from *G* to the station nearest to *H*, namely *F*.
(4) From co-ordinates of *G* and *F*, calculate the length and bearing of the line *GF*.

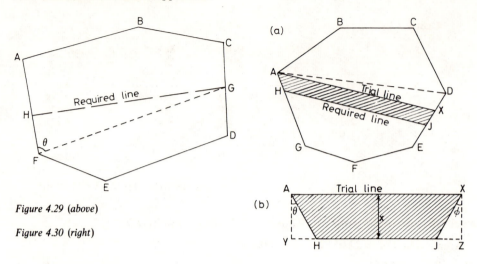

*Figure 4.29 (above)*

*Figure 4.30 (right)*

(5) Find the area of *GDEFG* and subtract this area from the required area to get the area of triangle *GFH*.
(6) Now area $GFH = 0.5HF \times FG \sin \theta$, distance *FG* is known from (4) above, and $\theta$ is the difference of the known bearings *FA* and *FG*, thus length *HF* is calculated.
(7) As the bearing *FH* = bearing *FA* (known), then the co-ordinates of *H* may be calculated.
(8) From co-ordinates of *G* and *H*, the length and bearing of *GH* are computed.

## 4.8.2 To cut off a required area by a line of given bearing

With reference to *Figure 4.30(a)*, it is required to fix line *HJ* of a given bearing, which divides the area *ABCDEFGA* into the required portions.

### Method

(1) From any station set off on the given bearing a trial line that cuts off approximately the required area, say *AX*.
(2) Compute the length and bearing of *AD* from the traverse co-ordinates.
(3) In triangle *ADX*, length and bearing *AD* are known, bearing *AX* is given and bearing *DX* = bearing *DE*; thus the three angles may be calculated and the area of the triangle found.
(4) From co-ordinates calculate the area *ABCDA*, thus total area *ABCDXA* is known.
(5) The difference between the above area and the area required to be cut off, is the area to be *added* or *subtracted* by a line *parallel* to the trial line *AX*. Assume this to be the trapezium *AXJHA* whose area is known together with the length and bearing of one side (*AX*) and the bearings of the other sides.
(6) With reference to *Figure 4.30(b)*, as the bearings of all the sides are known, the angles $\theta$ and $\phi$ are known. From which $YH = x \tan \theta$ and $JZ = x \tan \phi$, now:

Area of $AXJHA$ = area of rectangle $AXZYA$

$$- (\text{area of triangle } AHY + \text{area of triangle } XZJ)$$

$$= AX \times x - \left(\frac{x}{2} \times x \tan \theta + \frac{x}{2} \times x \tan \phi\right)$$

$$= AX \times x - \left[\frac{x^2}{2}(\tan \theta + \tan \phi)\right] \tag{4.9}$$

from which the value of $x$ may be found.

(7) Thus, knowing $x$ the distances $AH$ and $XJ$ can easily be calculated and used to set out the required line $HJ$.


## Worked examples

*Example 4.1.* The following table gives the co-ordinates of the sides of a traverse *ABCDEFA*.

| Side | ΔE (m) | ΔN (m) |
|------|--------|--------|
| AB | − 76.35 | −138.26 |
| BC | 145.12 | − 67.91 |
| CD | 20.97 | 109.82 |
| DE | 187.06 | 31.73 |
| EF | − 162.73 | 77.36 |
| FA | − 87.14 | − 25.24 |

It is apparent from these values that an error of 30 m has occurred, and is most likely to be in either *BC* or *EF*. Explain the reasons for these statements.

Tacheometric readings were taken from $A$ to a vertical staff at $D$. The telescope angle was 24° below horizontal and stadia readings of 1.737, 2.530 and 3.322 m were recorded. Use these readings to decide which length should be re-measured and also find the difference in level between stations $A$ and $D$ if the instrument height was 1.463 m above the station at $A$.  (LU)

Summing the above co-ordinates gives an error of $+26.93$ ($E$), $-12.5$ ($N$), the error vector being $(26.93^2 + 12.5^2)^{1/2} = 30$ m.

Thus, inspection of the above co-ordinates indicates the lines $BC$ or $EF$ as being the only possible sources of the error.

$$\text{Bearing of error vector} = \tan^{-1}\frac{26.9}{12.5} \approx \frac{2}{1}$$

$$\text{Bearing of } BC = \tan^{-1}\frac{145.12}{67.91} \approx \frac{2}{1}$$

$$\text{Bearing of } EF = \tan^{-1}\frac{162.73}{77.36} \approx \frac{2}{1}$$

Thus, the error could lie in either line as they are both parallel to the error vector. One must therefore utilize the tacheometric data as follows in order to isolate the line in question.

Distance $AD = 100S \cos^2 \theta$

$\qquad\qquad = 100 \times 1.585 \cos^2 24° = 132.3$ m

Distance $AD$ from co-ordinates $= (96.35^2 + 89.74^2)^{1/2} = 131.7$ m

Thus, the error of 30 m cannot be in the line $BC$ and must be in $EF$. An inspection of the co-ordinates indicates that $EF$ should be increased by 30 m.

Vertical height by tacheometry $= 132.3 \tan 24° = 58.90$ m

$\therefore$ Difference in level of $A$ and $D = 1.463 - 58.90 - 2.530 = 59.97$ m

*Example 4.2.* The following survey was carried out from the bottom of a shaft at $A$, along an existing tunnel to the bottom of a shaft at $E$.

| Line | WCB |  |  | Measured distance (m) | Remarks |
|------|-----|-----|-----|----------|---------|
|  | ° | ′ | ″ |  |  |
| AB | 70 | 30 | 00 | 150.00 | Rising 1 in 10 |
| BC | 0 | 00 | 00 | 200.50 | Level |
| CD | 154 | 12 | 00 | 250.00 | Level |
| DE | 90 | 00 | 00 | 400.56 | Falling 1 in 30 |

If the two shafts are to be connected by a straight tunnel, calculate the bearing $A$ to $E$ and the grade.

If a theodolite is set up at $A$ and backsighted to $B$, what is the value of the clockwise angle to be turned off, to give the line of the new tunnel?                     (KP)

Horizontal distance $AB = \dfrac{150}{(101)^{1/2}} \times 10 = 149.25$ m

Rise from $A$ to $B \qquad = 150 \div (101)^{1/2} = 14.92$ m

Fall from $D$ to $E \qquad = \dfrac{400.56}{(901)^{1/2}} = 13.34$ m

Horizontal distance $DE = \dfrac{400.56}{(901)^{1/2}} \times 30 = 400.34$ m

| Co-ordinates ($\Delta E, \Delta N$) | 0 | 0 | A |
|------|------|------|---|
| 149.25 $\genfrac{}{}{0pt}{}{\sin}{\cos}$ 70° 30′ 00″ | 140.69 | 49.82 | B |
| 200.50 due N | 0 | 200.50 | C |
| 250.00 $\genfrac{}{}{0pt}{}{\sin}{\cos}$ 154° 12′ 00″ | 108.81 | −225.08 | D |
| 400.34 due E | 400.34 | 0 | E |
| Total co-ords of $E$ | (E) 649.84 | (N)  25.24 |  |

∴ Tunnel is rising from $A$ to $E$ by $(14.92 - 13.34) = 1.58$ m

∴ Bearing $AE = \tan^{-1} \dfrac{+649.84}{+25.24} = 87° 47'$

Length $= 649.84/\sin 87° 47' = 652.33$ m

Grade $= 1.58$ in $652.33 = 1$ in $413$

Angle turned off $= BAE = (87° 47' - 70° 30') = 17° 17' 00''$

*Example 4.3.* A level railway is to be constructed from $A$ to $D$ in a straight line, passing through a large hill situated between $A$ and $D$. In order to speed the work, the tunnel is to be driven from both sides of the hill (*Figure 4.31*).

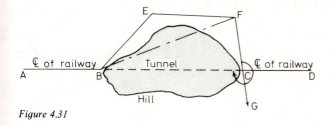

*Figure 4.31*

The centre-line has been established from $A$ to the foot of the hill at $B$ where the tunnel will commence, and it is now required to establish the centre-line on the other side of the hill at $C$, from which the tunnel will be driven back towards $B$.

To provide this data the following traverse was carried out around the hill:

| Side | Bearing ° ′ ″ | Horizontal distance (m) | Remarks |
|---|---|---|---|
| AB | 88  00  00 | — | Centre-line of railway |
| BE | 46  30  00 | 495.8 m | |
| EF | 90  00  00 | 350.0 m | |
| FG | 174  12  00 | — | Long sight past hill |

Calculate:

(1) The horizontal distance from $F$ along $FG$ to establish point $C$.

(2) The clockwise angle turned off from $CF$ to give the line of the reverse tunnel drivage.

(3) The horizontal length of tunnel to be driven.                    (KP)

*Find total co-ordinates of F relative to B*

|  | ΔE (m) | ΔN (m) | Stn |
|---|---|---|---|
| $495.8 \dfrac{\sin}{\cos} 46° 30' 00'' \rightarrow$ | E 359.6 | N 341.3 | BE |
| $350.0\ 90° 00' 00'' \rightarrow$ | E 350.0 | — | EF |
| Total co-ords of F | E 709.6 | N 341.3 | F |

$$\text{WCB of } BF = \tan^{-1} \frac{709.60}{341.30} = 64° 18' 48''$$

Distance $BF = 709.60/\sin 64° 18' 48'' = 787.42$ m

*Solve triangle BFC for the required data*

The bearings of all three sides of the triangle are known, from which the following values for the angles are obtained:

$FBC = 23° 41' 12''$
$BCF = 86° 12' 00''$
$CFB = 70° 06' 48''$
_____
$\qquad\quad 180° 00' 00''$   (check)

*By sine rule*

(a) $FC = \dfrac{BF \sin FBC}{\sin BCF} = \dfrac{787.42 \sin 23° 41' 12''}{\sin 86° 12' 00''} = 317.03$ m

(c) $BC = \dfrac{BF \sin CFB}{\sin BCF} = \dfrac{787.42 \sin 70° 06' 48''}{\sin 86° 12' 00''} = 742.10$ m

(b) $360° - BCF = 273° 48' 00''$

*Example 4.4.* The following Table shows details of a traverse *ABCDEFA*.

| Line | Length (m) | WCB | ΔE (m) | ΔN (m) |
|---|---|---|---|---|
| AB | 560.5 |  | 0 | − 560.5 |
| BC | 901.5 |  | 795.4 | − 424.3 |
| CD | 557.0 |  | − 243.0 | 501.2 |
| DE | 639.8 |  | 488.7 | 412.9 |
| EF | 679.5 | 293° 59′ |  |  |
| FA | 467.2 | 244° 42′ |  |  |

Adjust the traverse by the Bowditch method and determine the co-ordinates of the stations relative to *A* (0.0). What are the length and bearing of the line *BE*?    (LU)

*Complete the above Table of co-ordinates*

$$679.5 \, \frac{\sin}{\cos} \, 293° \, 59' \rightarrow$$

$$467.2 \, \frac{\sin}{\cos} \, 244° \, 42' \rightarrow$$

| Line | ΔE (m) | ΔN (m) |
|------|--------|--------|
| EF   | − 620.8 | + 276.2 |
| FA   | − 422.4 | − 199.7 |

Now refer *Table 4.8.*

TABLE 4.8

| Line | Lengths (m) | ΔE (m) | ΔN (m) | Corrected ΔE | Corrected ΔN | E | N | Stns |
|------|-------------|--------|--------|--------------|--------------|---|---|------|
| A    |             |        |        |              |              | 0.0 | 0.0 | A |
| AB   | 560.5 | 0 | − 560.5 | 0.3 | − 561.3 | 0.3 | − 561.3 | B |
| BC   | 901.5 | 795.4 | − 424.3 | 795.5 | − 425.7 | 796.2 | − 987.0 | C |
| CD   | 557.0 | − 243.0 | 501.2 | − 242.7 | 500.3 | 553.5 | − 486.7 | D |
| DE   | 639.8 | 488.7 | 412.9 | 489.0 | 411.9 | 1042.5 | − 74.8 | E |
| EF   | 679.5 | − 620.8 | 276.2 | − 620.4 | 275.2 | 422.1 | 200.4 | F |
| FA   | 467.2 | − 422.4 | − 199.7 | − 422.1 | − 200.4 | 0.0 Check | 0.0 Check | A |
| Sum  | 3805.5 | − 2.1 | 5.8 | 0.0 | 0.0 | | | |
| Correction to co-ordinates | | 2.1 | − 5.8 | | | | | |

The Bowditch corrections $(\delta E, \delta N)$ are computed as follows, and added algebraically to the co-ordinate differences, as shown in *Table 4.8.*

| Line | δE (m) | δN (m) |
|------|--------|--------|
|      | $\dfrac{2.1}{3805.5} \times 560.5$ giving | $\dfrac{-5.8}{3805.5} \times 560.5$ giving |
| AB | $K_2 \times 560.5 = 0.3$ | $K_1 \times 560.5 = -0.8$ |
| BC | $K_2 \times 901.5 = 0.5$ | $K_1 \times 901.5 = -1.4$ |
| CD | $K_2 \times 557.0 = 0.3$ | $K_1 \times 557.0 = -0.9$ |
| DE | $K_2 \times 639.8 = 0.3$ | $K_1 \times 639.8 = -1.0$ |
| EF | $K_2 \times 679.5 = 0.4$ | $K_1 \times 679.5 = -1.0$ |
| FA | $K_2 \times 467.2 = 0.3$ | $K_1 \times 467.2 = -0.7$ |
|    | Sum = 2.1 | Sum = − 5.8 |

To find the length and bearing of $BE$:

$$\Delta E = 1042.2 \quad \Delta N = 486.5$$

$$\therefore \text{ Bearing } BE = \tan^{-1} \frac{1042.2}{486.5} = 64° \ 59'$$

Length $BE = 1042.2/\sin 64° \ 59' = 1150.1 \text{ m}$

*Example 4.5.* In a quadrilateral $ABCD$ (*Figure 4.32*), the co-ordinates of the points, in metres, are as follows:

| Point | E | N |
|-------|------|---------|
| A | 0 | 0 |
| B | 0 | −893.8 |
| C | 634.8 | −728.8 |
| D | 1068.4 | 699.3 |

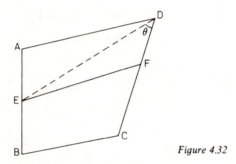

*Figure 4.32*

Find the area of the figure by calculation.

If $E$ is the mid-point of $AB$, find, either graphically or by calculation, the co-ordinates of a point $F$ on the line $CD$, such that the area $AEFD$ equals the area $EBCF$. (LU)

The above co-ordinates are total co-ordinates, therefore the appropriate rule is used.

| Stn | E | N | Difference of E | Sum of N | Double + | Area − |
|-----|------|--------|----------|---------|-----------|---------|
| A | 0 | 0 | 0 | −893.8 | | |
| B | 0 | −893.8 | −634.8 | −1622.6 | 1 030 026 | |
| C | 634.8 | −728.8 | −433.6 | −29.5 | | 12 791 |
| D | 1068.4 | 699.3 | 1068.4 | 699.3 | | 747 132 |
| A | 0 | 0 | | | | |

$\Sigma$   1 789 949

Area   894 974 m²

Rounding off the above values to the correct number of significant figures gives 895 000 m².

*To find the co-ordinates of F by calculation*

From co-ordinate geometry it is easily shown that the co-ordinates of $E$ are the mean of $A$ and $B$.

| Stn | E | N | Difference of E | Sum of N | Double + | Area − |
|---|---|---|---|---|---|---|
| A | 0 | 0 | 0 | − 446.9 | | |
| E | 0 | − 446.9 | − 1068.4 | 252.4 | | 269 700 |
| D | 1068.4 | 699.3 | 1068.4 | 699.3 | 747 100 | |

Σ  477 400

Area  238 700 m²

By co-ordinates, as above, area of triangle $AED$ is found.

$$\therefore \text{ Area of triangle } EDF = \frac{895\,000}{2} - 238\,700 = 208\,800 \text{ m}^2$$

*From co-ordinates*

$$\text{Bearing } ED = \tan^{-1} \frac{+1068.4}{+1146.2} = 42° \, 59'$$

$$\text{Length} = 1146.2 \cos 42° \, 59' = 1567.0 \text{ m}$$

$$\text{Bearing } DC = \tan^{-1} \frac{-433.6}{-1428.1} = 196° \, 54'$$

$$\therefore \theta = (42° \, 59' - 16° \, 54') = 26° \, 05'$$

Now: Area of triangle $EDF = \frac{1}{2} DE \times DF \sin \theta = 208\,800 \text{ m}^2$

$$\therefore DF = 208\,800/(0.5 \times 1567 \times \sin 26° \, 05') = 606 \text{ m}$$

Thus co-ordinates of $F$ relative to $D$ are

$$606 \, \genfrac{}{}{0pt}{}{\sin}{\cos} \, 196° \, 54' = -176.2(\Delta E) - 579.9(\Delta N)$$

$$\therefore \text{ Total co-ords of } F = 892.2 \text{ E } 119.4 \text{ N}$$

## Exercises

*(4.1)* In a closed traverse $ABCDEFA$ the angles and lengths of sides were measured, and, after the angles had been adjusted, the traverse sheet shown overleaf was prepared.

It became apparent on checking through the sheet that it contained mistakes. Rectify the sheet where necessary and then correct the co-ordinates by Bowditch's method. Hence, determine the co-ordinates of all the stations. The co-ordinates of $A$ are E − 235.5, N + 1070.0.

| Line | Length (m) | WCB ° ′ ″ | Reduced bearing ° ′ ″ | ΔE (m) | ΔN (m) |
|------|-----------|-----------|----------------------|--------|--------|
| AB | 355.52 | 58  30  00 | N 58  30  00 E | 303.13 | 185.75 |
| BC | 476.65 | 185  12  30 | S 84  47  30 W | −474.70 | − 43.27 |
| CD | 809.08 | 259  32  40 | S 79  32  40 W | −795.68 | −146.82 |
| DE | 671.18 | 344  35  40 | N 15  24  20 W | −647.08 | 178.30 |
| EF | 502.20 | 92  30  30 | S 87  30  30 E | 501.72 | − 21.83 |
| FA | 287.25 | 131  22  00 | S 48  38  00 E | 215.58 | −189.84 |

(Mistakes: Bearing BC to S 5° 12′ 30″ W, hence ΔE and ΔN interchange. ΔE and ΔN of DE interchanged. Bearing EF to S 87° 29′ 30″ E, giving new ΔN of − 21.97 m. Co-ords: (B) E 67.27, N 1255.18; (C) E 23.51, N 781.19; (D) E − 773.00, N 634.50; (E) E − 951.99, N 1281.69; (F) E − 450.78, N 1259.80)

(4.2) In a traverse ABCDEFG, the line BA is taken as the reference meridian. The co-ordinates of the sides AB, BC, CD, DE and EF are:

| Line | AB | BC | CD | DE | EF |
|------|-----|-----|-----|-----|-----|
| ΔN | −1190.0 | − 565.3 | 590.5 | 606.9 | 1017.2 |
| ΔE | 0 | 736.4 | 796.8 | − 468.0 | 370.4 |

If the bearing of FG is 284° 13′ and its length is 896.0 m, find the length and bearing of GA.　(LU)

(Answer: 947.8 m, 216° 45′)

(4.3) The following measurements were obtained when surveying a closed traverse ABCDEA:

| Line | EA | AB | BC | |
|------|-----|------|------|---|
| Length (m) | 793.7 | 1512.1 | 863.7 | |

| Included angles | DEA | EAB | ABC | BCD |
|-----------------|--------|---------|---------|---------|
| | 93° 14′ | 112° 36′ | 131° 42′ | 95° 43′ |

It was not possible to occupy D, but it could be observed from C and E. Calculate the angle CDE and the lengths CD and DE, taking DE as the datum, and assuming all the observations to be correct.　(LU)

(Answer: CDE = 96° 45′, DE = 1847.8 m, CD = 1502.0 m)

(4.4) An open traverse was run from A to E in order to obtain the length and bearing of the line AE which could not be measured direct, with the following results:

| Line | AB | BC | CD | DE |
|------|-----|-----|-----|-----|
| Length (m) | 1025 | 1087 | 925 | 1250 |
| WCB | 261° 41′ | 09° 06′ | 282° 22′ | 71° 31′ |

Find, by calculation, the required information.　(LU)

(Answer: 1620.0 m, 339° 46′)

(*4.5*) A traverse *ACDB* was surveyed by theodolite and chain. The lengths and bearings of the lines *AC*, *CD* and *DB* are given below.

| Line | AC | CD | DB |
|------|------|------|------|
| Length (m) | 480.6 | 292.0 | 448.1 |
| Bearing | 25° 19′ | 37° 53′ | 301° 00′ |

If the co-ordinates of *A* are $x = 0$, $y = 0$ and those of *B* are $x = 0$, $y = 897.05$, adjust the traverse and determine the co-ordinates of *C* and *D*. The co-ordinates of *A* and *B* must not be altered.                                                                (LU)

(*Answer:* Co-ord error: $x = 0.71$, $y = 1.41$, (*C*) $x = 205.2$, $y = 434.9$, (*D*) $x = 179.1$, $y = 230.8$) .

# 5

# Optical distance-measurement

Optical distance-measurement is rapidly being superseded by electromagnetic distance-measuring (EDM); therefore only those methods still viable will be dealt with here in any detail.

Vertical-staff tacheometry is viable largely because the equipment required (a theoeolite and levelling staff) is usually available.

The subtense bar has a unique place in the measuring processes available, affording high accuracy over short distances regardless of terrain conditions. Also, the concept of subtense measurement is still a valuable one.

In optical distance measurement, two basic techniques are used:

(1) Using a fixed parallactic angle and variable staff intercept.
(2) Using a fixed intercept and variable parallactic angle.

In both cases, the staff may be held either vertically or horizontally. In the UK, optical distance-measurement is generally termed *tacheometry*.

## 5.1 VERTICAL-STAFF TACHEOMETRY

The principle of this form of tacheometry, in which the parallactic angle $2\alpha$ remains fixed and the staff intercept $S$ varies with distance $D$, is shown in *Figure 5.1*. The parallactic angle is defined by the position of the stadia hairs, $c$ and $e$, each side of the main cross-hair $b$, then by similar triangles:

$$\frac{AB}{CE} = \frac{Ab}{ce}$$

put   $ce = i$

then   $D = (f/i)S = K_1 S$                  (5.1)

In modern telescopes $f$ and $i$ are so arranged that $K_1 = 100$.

Equation (5.1) is basically correct for horizontal sights taken with any modern instrument. The telescope will now be examined in more detail. In *Figure 5.2*, $f$ is the focal length of the object lens system, $d$ is the distance from the object lens to the centre of the instrument, $ce$ is the stadia interval, $i$ and $D$, the distance from the staff to the

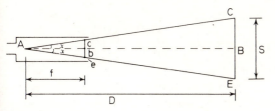

*Figure 5.1*

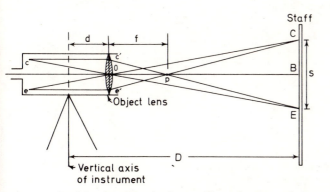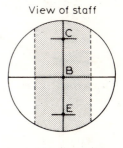

*Figure 5.2* On right, view through telescope illustrating the stadia lines at *C* and *E*

centre of the instrument, then by similar triangles:

$$\frac{Bp}{CE} = \frac{Op}{c'e'} \qquad \therefore \ Bp = S\left(\frac{f}{i}\right)$$

Now $\quad D = Bp + (f + d) = S(f/i) + (f + d)$

The value $(f + d)$ is called the *additive constant*, $K_2$, and $(f/i)$ is called the *multiplying constant*, $K_1$. Thus for horizontal sights:

$$D = K_1 S + K_2 \tag{5.2}$$

Tacheometry would have very little application if it was restricted to horizontal sights; thus the general formula will now be deduced. *Figure 5.3* illustrates an inclined sight.

By the sine rule in triangle *PCB*:

$$\frac{x_1}{\sin \alpha} = \frac{y \cot \alpha}{\sin [90° - (\theta + \alpha)]} = \frac{y \cot \alpha}{\cos (\theta + \alpha)}$$

Cross multiply.

$$x_1 \cos (\theta + \alpha) = y \cot \alpha \sin \alpha = \frac{y \cos \alpha \sin \alpha}{\sin \alpha} = y \cos \alpha$$

from which $\quad y = x_1 \cos \theta - x_1 \sin \theta \tan \alpha \tag{a}$

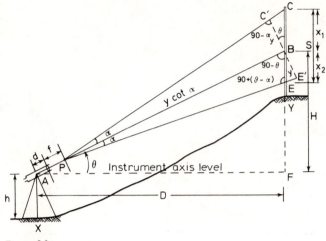

*Figure 5.3*

Similarly in triangle *PBE*

$$\frac{x_2}{\sin \alpha} = \frac{y \cot \alpha}{\sin [90° + (\theta - \alpha)]} = \frac{y \cot \alpha}{\cos (\theta - \alpha)}$$

$$x_2 \cos (\theta - \alpha) = y \cot \alpha \sin \alpha = y \cos \alpha$$

and $\qquad y = x_2 \cos \theta + x_2 \sin \theta \tan \alpha$ $\qquad\qquad$ (b)

Adding (a) and (b)

$$2y = (x_1 + x_2) \cos \theta - (x_1 - x_2) \sin \theta \tan \alpha$$

i.e. $\quad C'F' = S \cos \theta - (x_1 - x_2) \sin \theta \tan \alpha$ $\qquad\qquad$ (c)

The maximum value for $\sin \theta$ would be 0.707 ($\theta = 45°$) and for $\tan \alpha$, 0.005 ($\alpha = 1/200$), whilst for the majority of work in practice $x_1 \approx x_2$. Thus, the second term may be neglected for all but the steepest sights.

Now, from *Figure 5.3*

$$AB = K_1(C'E') + K_2 = K_1 S \cos \theta + K_2$$

$$\therefore \quad AF = D = AB \cos \theta = K_1 S \cos^2 \theta + K_2 \cos \theta \qquad\qquad \text{(d)}$$

Similarly $\quad FB = H = AB \sin \theta = K_1 S \cos \theta \sin \theta + K_2 \sin \theta$ $\qquad$ (e)

Alternatively $\quad H = D \tan \theta$ $\qquad\qquad\qquad\qquad\qquad\qquad\qquad$ (f)

In 1823, an additional anallactic lens was built into the telescope, which reduced all observations to the centre of the instrument and thus eliminated the additive constant $K_2$. All modern internal focusing telescopes, although not strictly anallactic may be regarded as so. Equations (d) and (e) therefore reduce to

$$D = K_1 S \cos^2 \theta \qquad\qquad\qquad\qquad\qquad\qquad\qquad\qquad\qquad \text{(g)}$$

and $\quad H = K_1 S \cos \theta \sin \theta$ $\qquad\qquad\qquad\qquad\qquad\qquad\qquad\qquad$ (h) but   $\cos\theta\sin\theta = \dfrac{1}{2}\sin 2\theta$ ⠀⠀⠀⠀⠀⠀⠀⠀⠀⠀⠀⠀⠀⠀⠀⠀⠀⠀⠀⠀⠀⠀(j)

$$\therefore\ H = \dfrac{1}{2}K_1 S \sin 2\theta$$

and where   $K_1 = 100$

$$D = 100S\cos^2\theta \tag{5.3}$$

$$H = 50S\sin 2\theta \tag{5.4}$$

With reference to *Figure 5.3* it can be seen that, given the reduced level of $X$ ($RL_X$), then the level of $Y$ is

$$RL_X + h_i + H - BY \tag{5.5}$$

If the sight had been from $Y$ to $X$ then a simple sketch as in *Figure 5.4* will serve to

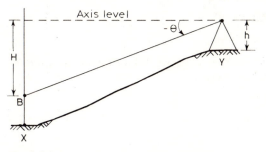

*Figure 5.4*

show that

$$RL_X = RL_Y + h_i - H - BX \tag{5.6}$$

where   $h_i$ = instrument height
⠀⠀⠀⠀⠀$BY$ or $BX$ = mid-staff reading

$H$ is positive when vertical angle is positive and *vice versa*.

Students should not attempt to commit formula (5.5) and equation (5.6) to memory, relying, if in doubt, on a quick sketch. Note that $H$ is always the vertical height from the centre of the transit axis to the mid-staff reading.

Thus, in general:

$$RL_X = (RL_Y + h_i) \pm H\text{—mid-staff reading} \tag{5.7}$$

where   $(RL_Y + h_i)$ = axis level

## 5.1.1 Inclined-staff tacheometry

Using the same equipment, the staff is fitted with a small sighting device to enable it to be held at right-angles to the line of sight (*Figure 5.5*):

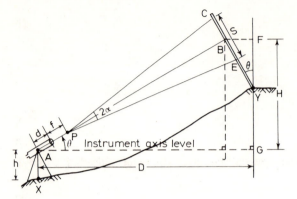

Figure 5.5

$$AB = K_1 S + K_2 \quad \text{and} \quad JG = BY \sin \theta$$

$$\therefore \ D = AJ + JG = AB \cos \theta + BY \sin \theta$$

$$= K_1 S \cos \theta + K_2 \cos \theta + BY \sin \theta$$

For an angle of depression term $BY \sin \theta$ would be negative. Where there is no additive constant and $K_1 = 100$, the expression becomes:

$$D = 100S \cos \theta \pm BY \sin \theta \tag{5.8}$$

Similarly $H = AB \sin \theta = K_1 S \sin \theta + K_2 \sin \theta$

when $\quad K_1 = 100 \quad$ and $\quad K_2 = 0,$

$$H = 100S \sin \theta \tag{5.9}$$

### 5.1.2 Measurements of tacheometric constants

Set up the instrument on fairly level ground giving horizontal sights to a series of pegs at known distances, $D$, from the instrument. Now, using the equation $D = K_1 S + K_2$ and substituting values for $D$ and $S$, the equations may be solved:

(1) Simultaneously in pairs and the mean taken.
(2) As a whole by the method of least squares. For example:

| Measured distance (m) | 30 | 60 | 90 | 120 | 150 | (D values) |
|---|---|---|---|---|---|---|
| Staff intercept (m) | 0.301 | 0.6 | 0.899 | 1.202 | 1.501 | (S values) |

from which $K_1 = 100$ and $K_2 = 0$ by either of the above methods.

### 5.1.3 Errors in staff holding

*Method* (1) Consider first the vertical-staff technique for which the basic equation is

$$D = K_1 S \cos^2 \theta$$

This equation is better expressed as

$$D = K_1 S \cos \theta_1 \cos \theta_2$$

where (in *Figure 5.3*) $\theta_1$ is the angle of inclination $BAF$ and $\theta_2$ is the angle $C'BC$, which, if there is error in the verticality of the staff $(\delta\theta_2)$, will be in error by the same amount.

Then, adopting the usual procedure for the treatment of small errors, the above expression is differentiated with respect to $\theta_2$, giving

$$\delta D = -K_1 S \cos \theta_1 \sin \theta_2 \, \delta\theta_2$$

$$\therefore \quad \frac{\delta D}{D} = \frac{-K_1 S \cos \theta_1 \sin \theta_2 \, \delta\theta_2}{K_1 S \cos \theta_1 \cos \theta_2} - \tan \theta_2 \, \delta\theta_2 \qquad (5.10)$$

Using the above expression the following table may be drawn up assuming $\theta_2 \approx \theta_1 = $ angle of inclination $= \theta$:

*Table 5.1* illustrates the following points:

Column 1   shows that if the staff is held reasonably plumb this source of error may be ignored.

Column 2   shows that the accuracy falls off rapidly as the angle of inclination increases.

Column 3   shows that where the staff is used carelessly, the accuracy is radically reduced, even on fairly level sights. It is obvious from this that all tacheometric staves should be fitted with a bubble and regularly checked.

**TABLE 5.1**

| $\theta$ | $\delta\theta_2 = 10'$ | $\delta\theta_2 = 1°$ | $\delta\theta_2 = 2°$ |
|---|---|---|---|
| 3° | 1/6670 | 1/1090 | 1/550 |
| 5° | 1/4000 | 1/650 | 1/330 |
| 10° | 1/1960 | 1/325 | 1/160 |
| 15° | 1/1280 | 1/215 | 1/110 |
| 20° | 1/940 | 1/160 | 1/80 |
| 25° | 1/740 | 1/120 | 1/60 |
| 30° | 1/600 | 1/100 | 1/50 |

*Method* (2) Consider now the inclined-staff technique for which the basic equation is $D = K_1 S \cos \theta$. Any error $(\delta\theta)$ in holding the staff normal to the line of sight will result in an increase in the stadia intercept $S$ to $S \sec \delta\theta$, as shown in *Figure 5.6*.

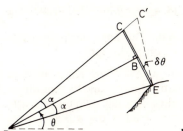

*Figure 5.6*

In triangle $C'CE$ as $\alpha$ is small, angle $C'CE \approx 90°$

$\therefore\ C'E = CE \sec \delta\theta = S \sec \delta\theta$

$\therefore$ Incorrect horizontal distance $D_e = K_1 S \sec \delta\theta \cos \theta$

Thus the error in the distance $= \delta D = D_e - D$

$$= K_1 \cos \theta\, S(\sec \delta\theta - 1)$$

$$\therefore\ \frac{dD}{D} = (\sec \delta\theta - 1) \tag{5.11}$$

Using the above expression *Table 5.2* may be drawn up indicating

(a) The error due to incorrect holding of the staff is entirely independent of the angle of inclination.
(b) Even gross errors of 2° may be considered to have a negligible effect.

**TABLE 5.2**

| $\delta\theta$ | $\delta D/D$ |
| --- | --- |
| 10′ | 1/238100 |
| 1° | 1/6560 |
| 2° | 1/1650 |
| 3° | 1/730 |

In comparing the two methods, it would appear that method (2) has all the advantages, including a simpler formula. However, method (1) is the one most used due to its easier method of staff holding. It is obvious though that where steep sights are involved, such as in open-cast mining or quarrying, method (2) should be used.

## 5.1.4 Errors in horizontal distances

(1) Careless staff holding, which has already been discussed.
(2) Error in reading the stadia intercept, which is immediately multiplied by 100 $(K_1)$, thereby making it significant. This source of error will increase with the length of sight. The obvious solution is to limit the length of sight to ensure good resolution of the graduations.
(3) Error in the determination of the instrument constants $K_1$ and $K_2$, resulting in an error in distance directly proportional to the error in the constant $K_1$ and directly as the error in $K_2$.
(4) Effect of differential refraction on the stadia intercept. This is minimised by keeping the lower reading 1 to 1.5 m above the ground.
(5) Random error in the measurement of the vertical angle. This has a negligible effect on the staff intercept and consequently on the horizontal distance.

In addition to the above sources of error, there are many others resulting from instrumental errors, failure to eliminate parallax, and natural errors due to high winds, heat shimmer, etc. The lack of statistical evidence makes it rather difficult to quote standards of accuracy; however, the usual treatment for small errors will give some basis for assessment.

Taking the equation for vertical staff tacheometry only, as in *Section 5.1.1*, and differentiating with respect to each of the sources of error, in turn gives

$$D = K_1 S \cos \theta_1 \cos \theta_2$$

thus,   $\delta D = S \cos \theta_1 \cos \theta_2 \, \delta K_1$

$$\therefore \quad \frac{\delta D}{D} = \frac{S \cos \theta_1 \cos \theta_2 \, \delta K_1}{S \cos \theta_1 \cos \theta_2 \, K_1} = \frac{\delta K_1}{K_1} \tag{5.12}$$

Similarly, differentiating with respect to $S$, $\theta_1$ and $\theta_2$, in turn gives

$$\delta D/D = \delta S/S$$

$$\delta D/D = -\tan \theta_1 \, \delta\theta_1$$

$$\delta D/D = -\tan \theta_2 \, \delta\theta_2$$

from the theory of errors, the sum effect of the above errors will give a fractional or proportional standard error (PSE) of

$$\delta D/D = \pm \left[ \left(\frac{\delta K_1}{K_1}\right)^2 + \left(\frac{\delta S}{S}\right)^2 + (\tan \theta_1 \, \delta\theta_1)^2 + (\tan \theta_2 \, \delta\theta_2)^2 \right]^{1/2} \tag{5.13}$$

Assuming now the following values: $D = 200$ m, $S = 2.015$ m, $\theta_1 = \theta_2 = 5°$, $\delta S = \pm(2^2 + 2^2)^{1/2} = \pm 3$ mm, $\delta K_1/K_2 = 1/1000$, $\delta\theta_1 = \pm 20''$ (error in vertical angle); $\delta\theta_2 = \pm 1°$ (error in staff holding).

*N.B.* $\delta\theta_1$ and $\delta\theta_2$ must always be expressed in radians (1 rad = 206 265'')

$$\delta D/200 = \pm \left[ (0.001)^2 + \left(\frac{0.003}{2.015}\right)^2 + (\tan 5°. 20'')^2 + (\tan 5°. 1°)^2 \right]^{1/2}$$

$$= \pm[(100 \times 10^{-8}) + (225 \times 10^{-8}) + \text{zero} + (234 \times 10^{-8})]^{1/2}$$

$$\therefore \quad \delta D = -0.48 \text{ m} \quad \text{and} \quad \delta D/D = 1 \text{ in } 420$$

It is obvious that the most serious sources of error lie in careless staff holding and stadia intercept error, the error in vertical angles being negligible. Reading to the nearest 10 mm gives a maximum error of $\pm 5$ mm and an average error of $\pm 2.5$ mm. The average error in the value of the inercept would therefore be $(2.5^2 + 2.5^2)^{1/2} = \pm 3.5$ mm. Using this value an accuracy of 1 in 400 is obtained, but this will fall off rapidly, with increase in distance and elevation. For most engineers, an accuracy of 1 in 250 is more realistic under normal field conditions, considering that the work is not generally carried out by expert users.

## 5.1.5 Errors in elevations

The main sources of error in elevation are (1) error in vertical angles, (2) additional errors arising from errors in the computed distance. *Figure 5.7* clearly shows that whilst the error resulting from (1) remains fairly constant, that resulting from (2) increases with increased elevation.

$$H = D \tan \theta$$

$$\therefore \quad \delta H = \delta D \tan \theta$$

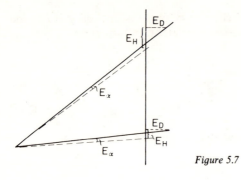

*Figure 5.7*

$$\delta H = D \sec^2 \theta \, \delta\theta$$

$$\therefore \quad \delta H = \pm [(\delta D \tan \theta)^2 + (D \sec^2 \theta \, \delta\theta)^2]^{1/2}$$

$$= \pm [(0.48 \tan 5°)^2 + (200 \sec^2 5° \times 20'' \sin 1'')^2]^{1/2}$$

$$= \pm 0.046 \text{ m}$$

This result indicates that elevations need be quoted only to the nearest 10 mm.

Accuracies of 1 in 1000 may be still achieved in tacheometric traversing, due to the compensating effect of accidental errors, reciprocal observation of the lines and a general increase in care taken.


## 5.2 APPLICATION

The method is easy to use in the field but, unless a direct-reading tacheometer (refer *Section 5.4.1*) is used, the resultant computation for many 'spot-shots' can be extremely tedious, even with the use of a computer program.

The very low order of accuracy and its short range limit its application to detail surveys in rural areas or contouring.


### 5.2.1 Detail survey

The method of fixing the position of topographic detail (in three dimensions) is by *polar co-ordinates.*

The theodolite is set up at a control station *A* (*Figure 5.8*) and oriented to any other control station (RO) with the horizontal circle set to 0°00′. Thereafter the bearing (relative to $A - $ RO) and horizontal length to each point of detail (*P1, P2, P3*, etc.) are obtained by observing the stadia readings on a staff held there, the horizontal circle reading ($\phi_1, \phi_2, \phi_3$, etc.) and the vertical angle. The cross-hair reading is also required to compute the reduced level of the points.

The field data is booked as shown in *Table 5.3* and reduced using equations (5.3), (5.4) and (5.7). Note that the angles are required to the nearest minute of arc only.

It is worth noting that the staff-man should be the most experienced member of the survey party who would appreciate the error sources, the limited accuracy available and thus the best and most economic staff positions required.

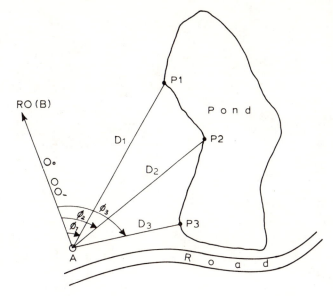

*Figure 5.8*

## 5.2.2 Contouring

Contouring is carried out in exactly the same manner as above, but with many more spot-shots along each radial arm (*Figure 5.9*). The arms are turned off at regular angular intervals, with the staff-man obtaining levels at regular paced intervals along each arm and at each distinct change in gradient. Subsequent computation of the field data will fix the position and level of each point along each arm, which may then be interpolated for contours.

## 5.3 SUBTENSE TACHEOMETRY

This method uses a horizontal subtense bar with targets at each end precisely 2 m apart. Although the bar is of steel construction, the targets are connected to an invar wire in such a way as to compensate for temperature changes. The bar can be set up horizontally on a normal theodolite tribrach and set at 90° to the line of sight by means of a small sighting device at its centre (*Figure 5.10*).

### 5.3.1 Principle of operation

The principle is illustrated by *Figure 5.11*. Regardless of the elevation, the angle $\theta$ subtended by the bar is measured in the horizontal plane by the theodolite. The horizontal distance $TB$ is then given by

$$D = b/2 \cot \theta/2 \qquad\qquad (5.14)$$

$$= \cot \theta/2 \quad \text{when} \quad b = 2\ \text{m}$$

**TABLE 5.3**

| At station | A | | Stn level (RL) | 30.48 m OD |
|---|---|---|---|---|
| Grid ref | E 400, N 300 | | Ht of inst ($h_i$) | 1.42 m |
| Weather | Cloudy, cool | | Axis level (RL + $h_i$) | 31.90 m (Ax) |

Survey __Canbury Park__
Surveyor __J. SMITH__
Date __12.12.83__

| Staff point | Angles observed Horizontal | Vertical circle | Vertical angle ±θ | Staff readings | Staff intercept S | Horizontal distance $K.S.\cos^2\theta$ D | Vertical height $\frac{K}{2}.S.\sin 2\theta$ ±H | Reduced level Ax ± H − m | Remarks |
|---|---|---|---|---|---|---|---|---|---|
| | ° ′ | ° ′ | ° ′ | m | | | | | |
| R0 | 0 00 | | | m | | | | | Station B |
| P1 | 48 12 | 95 20 | − 5 20 | 1.942 / m 1.404 / 0.866 | 1.076 | 106.67 | −9.96 | 20.54 | Edge of pond |
| P2 | 80 02 | 93 40 | − 3 40 | 0.998 / m 0.640 / 0.281 | 0.717 | 71.41 | −4.58 | 26.68 | Edge of pond |
| P3 | 107 56 | 83 20 | + 6 40 | 1.610 / m 1.216 / 0.822 | 0.788 | 77.74 | +9.09 | 39.77 | Edge of pond |

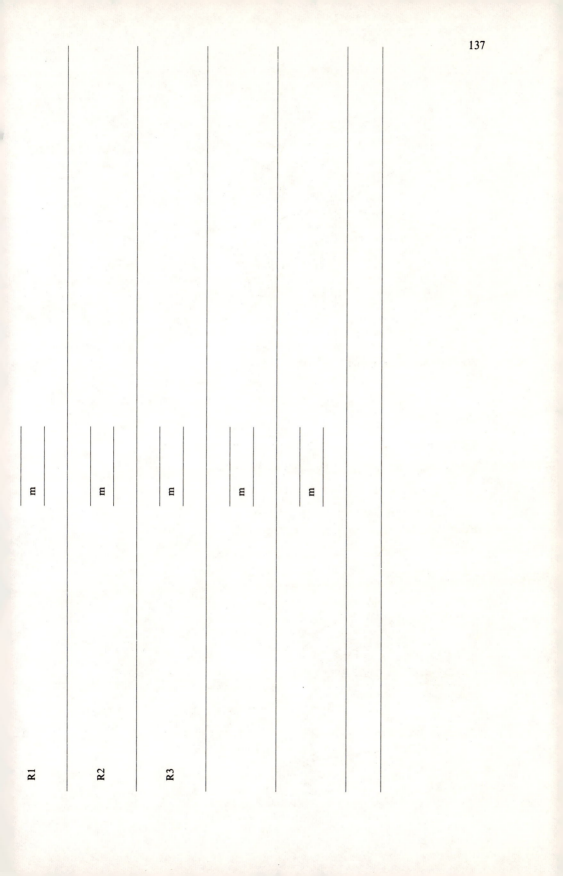

R1

R2

R3

m

m

m

m

m

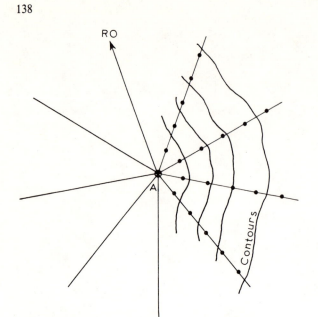

RO

A

Contours

*Figure 5.9* Radiation method

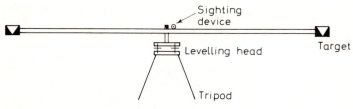

Sighting
device

Levelling head

Target

Tripod

*Figure 5.10*

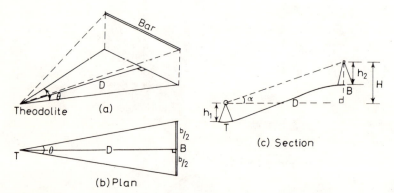

Bar

D

θ

Theodolite (a)

T
θ
D
B
b/2
b/2

(b) Plan

h₂

H

α

h₁

T

D

d

B

(c) Section

*Figure 5.11*

The vertical distance is given by

$$H = D \tan \alpha \qquad (5.15)$$

and the level of $B$ relative to $T$, would therefore be

Level of $B$ = level of $T + h_1 + H - h_2$

showing that in the computation of levels, one would require the instrument heights.

## Errors

The three sources of error in the distance $D$ are

(1) Variation in the length of the subtense bar.
(2) Error in setting the bar at 90° to the line of sight, and horizontally.
(3) Error in the measurement of the subtense angle.

To simplify the differentiation of each variable the basic formula is reduced to a form as follows:

$D = b/2 \cot \theta/2$, but since $\theta/2$ is very small
$\tan \theta/2 \approx \theta/2$ rad, thus $\cot \theta/2 = 2/\theta$
$\therefore \ D = (b/2)(2/\theta) = b/\theta$

It can be shown that the error in this approximation is roughly 1 in 3 $D^2$ and should therefore never be used for the reduction of sights; e.g. when $D = 40$ m, $b/\theta$ is accurate to only 1 in 4800.

## (1) Error in bar length

$$D = b/\theta$$

$$\therefore \ \delta D = \frac{\delta b}{\theta} \qquad \text{and} \qquad \frac{\delta D}{D} = \frac{\delta b}{\theta} \frac{\theta}{b}$$

Thus $\dfrac{\delta D}{D} = \dfrac{\delta b}{b}$ \qquad (5.16)

Manufacturers of the various subtense bars claim a value of 1/100 000 for $\delta b/b$ due to a 20°C change in temperature. This source of error may therefore be ignored.

## (2) Error in bar setting

Failure to align the bar at 90° to the line of sight results in the length $AC$ being reduced to $A'C' \approx b \cos \phi$ (*Figure 5.12*). Misalignment in the vertical plane, however, shows $A'C' = b \cos \phi$. Thus, the error in the bar length $= b - b \cos \phi$ in both cases

i.e. $\delta b = b(1 - \cos \phi)$, then from equation (5.16) above
$\delta D/D = \delta b/b = (1 - \cos \phi)$
but $\cos \phi = 1 - \phi^2/2! + \phi^4/4! \dots$
$\therefore \ \delta D/D = \phi^2/2$ \qquad (5.17)

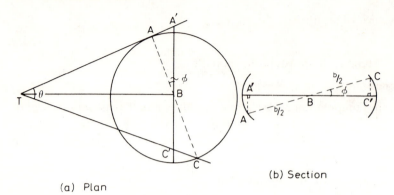

(b) Section

(a) Plan

*Figure 5.12*

If $\delta D/D$ is not to exceed 1/20 000 then

$$\phi = \left(\frac{2}{20\,000}\right)^{1/2} = 1/100 \text{ rad} \approx 0° \, 34'$$

Alignment to this accuracy is easily obtained by using the standard sighting devices. This source of error may therefore be ignored.

### (3) *Error in the measurement of the subtense angle*

$$D = \frac{b}{\theta} \qquad \therefore \; \delta D = \frac{-b}{\theta^2}\,\delta\theta = \frac{-b}{\theta}\frac{\delta\theta}{\theta} = -D\frac{\delta\theta}{\theta}$$

$$\therefore \; \frac{\delta D}{D} = \frac{\delta\theta}{\theta} \tag{5.18}$$

Using the above relationship *Table 5.4* may de deduced, assuming a 2-m bar and an error of $\pm 1''$ in the measurement of $\theta$, illustrating that the accuracy falls off rapidly with increase in distance. By further manipulation of the above equation it can be shown that the error in $D$ varies as the square of the distance:

$$\delta D = -(b/\theta^2)\,\delta\theta \qquad \text{but } \theta^2 = b^2/D^2$$

$$\therefore \; \delta D = (D^2/b)\,\delta\theta \tag{5.19}$$

Thus, an error of $\pm 1''$ produces four times the error at 80 m than it does at 40 m. This can be further clarified from *Table 5.4* where $40/10\,000 = 40$ mm, $80/5000 = 16$ mm.

To achieve a PSE of 1/10 000 the distance must be limited to 40 m and an accuracy of

**TABLE 5.4**

| $D$ (m) | 20 | 40 | 60 | 80 | 100 |
|---|---|---|---|---|---|
| $\delta D/D$ | 1 in 20 626 | 1 in 10 313 | 1 in 6875 | 1 in 5106 | 1 in 4125 |

$\pm 1''$ attained in the measurement of the angle. This is possible only with a $1''$-reading theodolite.

Research has proved that the subtense angle should be measured at least eight times to achieve the necessary accuracy. As a $1''$ instrument would be used, there is no need to change face between observations to eliminate instrumental errors, as each end of the bar is at the same elevation. However, to eliminate graduation errors they should be observed on different parts of the horizontal circle.

## 5.3.2 Serial measurement

In order to extend the range of the equipment and still have a comparable accuracy, the method called *serial measurement* may be used (*Figure 5.13*). The error in one sub-section $d$ from equation (5.19) is

$$\delta d = (d^2/b)\,\delta\theta$$

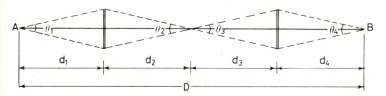

Figure 5.13

From the theory of errors, the standard error in the total distance is

$$\delta D_n = [(\delta d_1)^2 + (\delta d_2)^2 + (\delta d_3)^2 + \ldots + (\delta d_n)^2]^{1/2}$$

Assuming $d_1 = d_2 = d_n$ and $\theta_1 = \theta_2 = \theta_n$ then

$$\delta D_n = \delta d : n^{1/2} = \frac{d^2\,\delta\theta\,n^{1/2}}{b}$$

Now $D = n.d$ $\therefore$ $d^2 = D^2/n^2$ which on substituting above gives

$$\delta D_n = \frac{D^2\,\delta\theta n^{1/2}}{bn^2} = \frac{D^2\,\delta\theta}{b(n^3)^{1/2}} \qquad (5.20)$$

but from equation (5.19) $\delta D = (D^2/b)\,\delta\theta$

$$\therefore \ \delta D_n = \delta D/(n^3)^{1/2} \qquad (5.21)$$

Splitting the line into two bays only ($n = 2$), and assuming a standard error of $\pm 1''$, the maximum distance at which the accuracy of 1 in 10 000 is maintained can be found by expressing equation (5.20) as follows:

$$\frac{\delta D_n}{D} = \frac{D\,\delta\theta}{b(n^3)^{1/2}}$$

N.B. The small angles in the equation ($\delta\theta$) must *always* be expressed in radians (1 rad = 206 265''). Thus:

$$\frac{1}{10\ 000} = \frac{D \times 1''}{2 \times (2^3)^{1/2} \times 206\ 265}$$

$$\therefore\ D = 117\ \text{m}$$

### 5.3.3 Auxiliary base measurement

Beyond two bays serial measurement becomes uneconomical and the *auxiliary base* technique should be adopted for distances in excess of 117 m. In *Figure 5.14* the distance required is $AB$. The auxiliary base $BC$ is established at 90° to $AB$ and measured

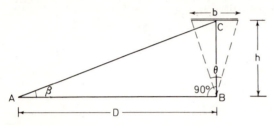

Figure 5.14

with the subtense bar at $C$. Measurement of the angle $\beta$ at $A$, enables $AB$ to be obtained from $AB = D = h \cot \beta$.

The following analysis of the errors enables optimum distances for $h$ and $D$ to be calculated.

$$h \approx b/\theta \qquad \text{and} \qquad D \approx h/\beta \approx b/\theta \cdot \beta$$

Differentiating with respect to $\theta$ and $\beta$ in turn gives

$$\delta D = -(b/\theta^2 \beta)\,\delta\theta \qquad \text{and} \qquad \delta D = -(b/\theta\beta^2)\,\delta\beta$$

$$\therefore\ \delta D = \pm \left[ \frac{b^2}{\theta^4 \beta^2}\,\delta\theta^2 + \frac{b^2}{\theta^2 \beta^4}\,\delta\beta^2 \right]^{1/2}$$

$$= \pm \frac{b}{\theta \cdot \beta} \left[ \left(\frac{\delta\theta}{\theta}\right)^2 + \left(\frac{\delta\beta}{\beta}\right)^2 \right]^{1/2}$$

$$= \pm D \left[ \left(\frac{\delta\theta}{\theta}\right)^2 + \left(\frac{\delta\beta}{\beta}\right)^2 \right]^{1/2}$$

It is logical to assume that the proportional standard error in each angle would be the same, thus

$$\frac{\delta D}{D} \pm \left[ 2\left(\frac{\delta\theta}{\theta}\right)^2 \right]^{1/2} \tag{5.22}$$

$$\text{or} \qquad \pm \left[ 2\left(\frac{\delta\beta}{\beta}\right)^2 \right]^{1/2} \tag{5.23}$$

Now as $\theta \approx b/h$ and $\beta \approx h/D$ formulae (5.22) and (5.23) may be shown respectively as follows:

$$\frac{\delta D}{D} = \pm \left[ \frac{2h^2 \, \delta\theta^2}{b^2} \right]^{1/2} = \pm \frac{h \, \delta\theta}{b} (2)^{1/2} \tag{5.24}$$

and $\quad \dfrac{\delta D}{D} = \pm \dfrac{D \, \delta\beta}{h} (2)^{1/2} \tag{5.25}$

Assuming once again that the required accuracy is 1 in 10 000, and the standard error is $\pm 1''$, using a 2-m bar.

$$\frac{1}{10\ 000} = \frac{h}{2 \times 206\ 265} (2)^{1/2}$$

from which $\quad h = 29$ m

Now using the above data

$$\frac{1}{10\ 000} = \frac{D}{29 \times 206\ 265} (2)^{1/2}$$

from which $\quad D = 425$ m

Finally, there will be error in the setting out of the 90° angle at $B$; however, as this source of error can be shown to be proportional to the cotangent, the error will be zero, if the angle is 90°.

Further to the above, the length of $D$ may be extended by having the auxiliary base at the centre (as shown in *Figure 5.15*), then:

$$D = D_1 + D_2 = \frac{b}{\theta\beta_1} + \frac{b}{\theta\beta_2}$$

as shown previously.

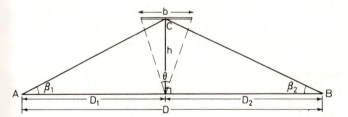

*Figure 5.15*

$D = b/\theta(1/\beta_1 + 1/\beta_2)$ which on differentiating with respect to $\theta$, $\beta_1$ and $\beta_2$ gives

$$\delta D = \frac{-b}{\theta^2} \left( \frac{1}{\beta_1} + \frac{1}{\beta_2} \right) \delta\theta; \quad \delta D = -\frac{b \, \delta\beta_1}{\theta\beta_1^2}; \quad \delta D = -\frac{b \, \delta\beta_2}{\theta\beta_2^2}$$

$$\therefore \ \delta D = \pm \left[ \frac{b^2}{\theta^4} \left( \frac{1}{\beta_1} + \frac{1}{\beta_2} \right)^2 \delta\theta^2 + \frac{b^2 \, \delta\beta_1^2}{\theta^2 \beta_1^4} + \frac{b^2 \, \delta\beta_2^2}{\theta^2 \beta_2^4} \right]^{1/2}$$

It is logical to assume that the angular errors throughout would be equal

$$\therefore \ \delta D = \pm \left[ \frac{b^2 2^2 \, \delta\theta^2}{\theta^6} + \frac{b^2 \, \delta\theta^2}{\theta^6} + \frac{b^2 \, \delta\theta^2}{\theta^6} \right]^{1/2}$$

$$= \frac{b \, \delta\theta}{\theta^3} (6)^{1/2} \tag{5.26}$$

Equation (5.26) may be written as

$$\frac{b \, \delta\theta}{\theta^2 \, \theta} \left( 4 \times \frac{3}{2} \right)^{1/2} = \frac{2b \, \delta\theta}{\theta^2 \, \theta} \left( \frac{3}{2} \right)^{1/2}$$

now assuming

$$D_1 \approx D_2, \qquad D = \frac{2b}{\theta\beta} = \frac{2b}{\theta^2} \quad \text{then it becomes}$$

$$\delta D = D \frac{\delta\theta}{\theta} \left( \frac{3}{2} \right)^{1/2} \qquad \text{and as} \quad \theta = \frac{b}{h}$$

$$\frac{\delta D}{D} = \frac{\delta\theta h}{b} \left( \frac{3}{2} \right)^{1/2} \tag{5.27}$$

Using equation (5.27) and assuming $\delta D/D = 1$ in 10 000 and $\delta\theta = \pm 1''$,

then  $h = 34$ m

Similarly, from equation (5.26), i.e.

$$\delta D = \frac{b \, \delta\theta}{\theta^3} (6)^{1/2} \qquad \text{as} \quad D = \frac{2b}{\theta^2} \qquad \text{then} \quad \theta^3 = \left( \frac{2b}{D} \right)^{3/2}$$

which on substitution gives

$$\delta D = \frac{b \, \delta\theta 6^{1/2} D^{3/2}}{(8b^3)^{1/2}}$$

$$\therefore \ \frac{\delta D}{D} = \frac{b \, \delta\theta 6^{1/2} D^{1/2}}{(8b^3)^{1/2}}$$

Substituting the same values as above:

$$\frac{1}{10\,000} = \frac{2 \times 1'' \times 6^{1/2} D^{1/2}}{(8 \times 2^3)^{1/2} \times 206\,265}$$

$$\therefore \ D = \left( \frac{8 \times 206\,265}{20\,000 \times 6^{1/2}} \right)^2 = 1132 \text{ m}$$

This basic approach to small errors can be used to build up different procedures which maintain the same accuracy but increase the distance measured. For instance, *Figure 5.16* indicates a procedure which enables a distance of 3400 m ($D$) to be measured to an accuracy of 1 in 10 000, provided that $h_1 = 25$ m, $h_2 = 280$ m and $\delta\theta = \pm 1''$.

These various techniques may also be incorporated into traversing using three tripod equipment, and for sights up to, say, 400 m, the procedure may be as illustrated in *Figure 5.17*.

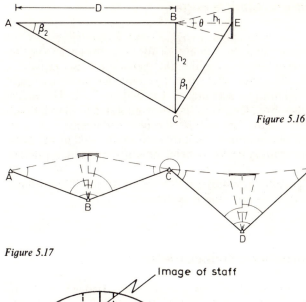

*Figure 5.16*

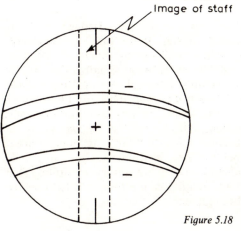

*Figure 5.17*

Image of staff

*Figure 5.18*

## 5.4 FURTHER OPTICAL DISTANCE-MEASURING EQUIPMENT

### 5.4.1 Direct-reading tacheometers

*Direct-reading tacheometers*, or *self-reducing tacheometers* as they are also called, have curved lines replacing the conventional stadia lines. *Figure 5.18* illustrates one particular make in which the outer lines are curves to the function $\cos^2 \theta$ and the inner curves are to the function $\sin \theta \cos \theta$. Thus the outer curve staff intercept is not just $S$ but $S \cos^2 \theta$, hence one need only multiply the intercept reading by $K_1 = 100$ to obtain the horizontal distance. Similarly, the inner curve staff intercept is $S \sin \theta \cos \theta$, and need only be multiplied by $K_1$ to produce the vertical height $H$. The separation of the curves varies with variation in the vertical angle.

There are other makes of instrument which have different methods of solution; however, the objective remains the same, i.e., to eliminate computation. It should be noted that there is no improvement in accuracy.

### 5.4.2 Distance-measuring optical wedge

The *distance-measuring wedge* is an achromatic wedge of glass accurately ground to refract rays of light by an amount equal to 1 in 100 of the *slant* distance from the instrument to the staff. It attaches to the object end of the theodolite telescope (with a counterweight at the eyepiece) and is used with a specially-graduated horizontal staff, to measure distances of up to 150 m to an accuracy of 1 in 5000 to 7500. The wedge covers only the mid-section of the object lens; thus the horizontal staff is viewed direct through the uncovered portions of the lens, whilst the deflected image is viewed through the wedge. The amount of deviation is read direct from the staff as the slope distance to the nearest 0.1 m, finer setting to 0.01 m being obtained by an incorporated parallel-plate micrometer.

This attachment is difficult to use in hot, sunny conditions, due to heat shimmer and refraction. It is rapidly being superseded by EDM equipment.

### 5.4.3 Horizontal-staff precision tacheometer

The *horizontal-staff tacheometer* utilizes an extension of the above wedge principle to produce the horizontal distance from instrument to staff. In this case two achromatic wedges are used, each ground to refract rays of light by an amount equal to 1/200 of the slant distance. When the two wedges are together, the telescope is horizontal and the amount of deviation is 1/100. As the telescope is rotated through a vertical angle of, say, $\theta$, the wedges move in opposite directions through the same angle and the resultant displacement vector is reduced in proportion to $\cos \theta$. The horizontal distance may then be read direct from a special horizontal staff. *Figure 5.19* shows *AB* is the displacement equivalent to the horizontal distance required. With a maximum range of 250 m, an accuracy of 1 in 10 000 is claimed for these instruments.

These instruments are also extremely difficult to use in hot, sunny conditions. They are very idiosyncratic and require careful and regular calibration. It is felt they are rapidly being rendered obsolete by EDM equipment.

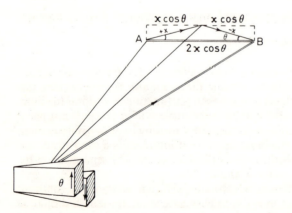

*Figure 5.19* Wedges rotating in opposite directions through angle of elevation $\theta$

### 5.4.4 Vertical-staff precision tacheometer

The *vertical-staff tacheometer*, as produced by Kern and named the Kern *DK—RV*, has a moveable diaphragm which varies with the inclination of the telescope, the amount of variation being controlled by a gear-and-cam mechanism. It is used with a specially-graduated vertical staff giving horizontal distances to an accuracy of 1 in 5000 over a maximum range of 150 m. *Figure 5.20* illustrates a portion of the special staff as viewed through the instrument. By rotating the telescope in the vertical plane, the horizontal

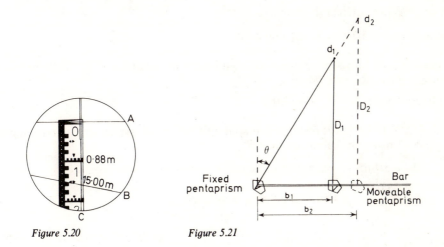

Figure 5.20                    Figure 5.21

reticule *A* is made to bisect the zero wedge. Rotation of the instrument in azimuth is carried out until the sloping reticule *B* bisects a small circular dot on the left-hand scale. The instrument now reads as follows:

Reticule *B*              = 15.00 m
Vertical reticule *C*  =   0.88 m
_____
Horizontal distance = 15.88 m

The same comments apply to this instrument as to the horizontal-staff precision tacheometer.

### 5.4.5 Zeiss BRT 006

The Zeiss *BRT 006* is an improved version of the well-known Teletop, the principle of which is illustrated in *Figure 5.21*. Using a constant parallactic angle $\theta$ and a variable base along the bar, the distance *D* is obtained by shifting the moveable prism along the bar until the two images of $d_1$ are coincident. Slant and/or horizontal distance is then read direct from the bar. The error in measurement is directly proportional to the distance, the normal range being limited to 60 m, accurate to 1 in 1600, but by using special targets the range can be extended to 180 m. The instrument is ideal for detail surveying in busy urban areas and, used in conjunction with the Plotting Table Karti attachment, affords direct semi-automatic plotting of points to an accuracy of $\pm 0.1$ mm.

This instrument provides a unique solution to the problem of accurate fixing of position using remote methods and has a special place in the catalogue of equipment available to the surveyor.

## Worked examples

*Example 5.1.* Consider a distance of 500 m. To what accuracy would it be measured using a 2-m bar and assuming a standard error of $\pm 1''$ in the subtense angle?

$$\theta = 2/500 \text{ rad} = 0.004 \times 206\,265 = 825''$$

now as   $\delta\theta/\theta = \delta D/D = 1/825$ (606 mm)

*Example 5.2.* If the above distance was required to an accuracy of 1 in 1000, to what accuracy must the angle be measured?

$$1/1000 = \delta\theta''/825 \quad \therefore \quad \delta\theta = \pm 0.8''$$

*Example 5.3.* If the above distance was split into two equal bays what accuracy might be expected using the same equipment?

$$\delta D_2 = \delta D/n^{3/2} \quad \text{where from e.g.1.} \quad \delta D = 606 \text{ mm} \quad \text{and} \quad n = 2$$

$$\therefore \quad \delta D_2 = \frac{606}{8^{1/2}} = 214 \text{ mm in } 500 \text{ m} = 1 \text{ in } 2338$$

*Example 5.4.* How many bays would be required to increase the accuracy to 1 in 10 000?

$$\delta D_n = \frac{500 \text{ m}}{10\,000} = 50 \text{ mm}$$

$$\therefore \quad 50 = \frac{606}{n^{3/2}} \quad \text{from which} \quad n = 5.3 \text{ bays}$$

*Example 5.5.* As the use of five bays would be uneconomical, what length of auxiliary base would give the required accuracy?

$$\frac{\delta D}{D} = \frac{D \, \delta\beta}{h} 2^{1/2}$$

$$\therefore \quad \frac{1}{10\,000} = \frac{500 \times 1'' \times (2)^{1/2}}{206\,265h} \quad \text{from which} \quad h = 34 \text{ m}$$

*Example 5.6.* A theodolite has a tacheometric constant of 100 and an additive constant of zero. The centre reading on a vertical staff held on a point $B$ was 2.292 m when sighted from $A$. If the vertical angle was $+25°$ and the horizontal distance $AB$ 190.326 m, calculate the other staff readings and thus show that the two intercept intervals are not equal. Using these values calculate the level of $B$ if $A$ was 37.95 m and the height of the instrument 1.35 m.                                          (LU)

From basic equation (5.3)  $CD = 100S \cos^2 \theta$

$$190.326 = 100S \cos^2 25°$$

$$\therefore \quad S = 2.316 \text{ m}$$

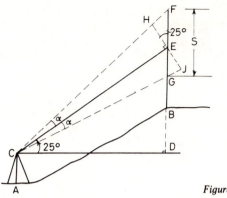

Figure 5.22

From *Figure 5.22*:   $HJ = S \cos 25° = 2.1$ m

Inclined distance    $CE = CD \sec 25° = 210$ m

$$\therefore\ 2\alpha = \frac{2.1}{210} \text{ rad} = 0° \ 34' \ 23''$$

$$\therefore\ \alpha = 0° \ 17' \ 11''$$

Now, by reference to *Figure 5.22*:

$DG = CD \tan (25° - \alpha) = 87.594$
$DE = CD \tan 25° \qquad = 88.749$
$DF = CD \tan (25° + \alpha) = 89.910$

It can be seen that the stadia intervals are

$$\left.\begin{array}{l} GE = S_1 = 1.155 \\ EF = S_2 = 1.161 \end{array}\right\} = 2.316 \text{ (check)}$$

from which it is obvious that the

upper reading $= (2.292 + 1.161) = 3.453$
lower reading $= (2.292 - 1.155) = 1.137$

Vertical height   $DE = H = CD \tan 25° = 88.749$ (as above)

$$\therefore\ \text{Level of } B = 37.95 + 1.35 + 88.749 - 2.292 = 125.757 \text{ m}$$

*Example 5.7.* The following observations were taken with a tacheometer, having constants of 100 and zero, from a point $A$ to $B$ and $C$. The distance $BC$ was measured as 157 m. Assuming the ground to be a plane within the triangle $ABC$, calculate the volume of filling required to make the area level with the highest point, assuming the sides to be supported by vertical concrete walls. Height of instrument was 1.4 m, the staff held vertically.
(LU)

| At | To | Staff readings (m) | | | Vertical angle | |
|----|----|----|----|----|----|----|
| $A$ | $B$ | 1.48, | 2.73, | 3.98 | $+7°$ | $36'$ |
|  | $C$ | 2.08, | 2.82, | 3.56 | $-5°$ | $24'$ |

Horizontal distance $AB = 100 \times S \cos^2 \theta$
$$= 100 \times 2.50 \cos^2 7° \ 36' = 246 \text{ m}$$
Vertical distance $AB = 246 \tan 7° \ 36' = +32.8 \text{ m}$

## Similarly

Horizontal distance $AC = 148 \cos^2 5° \ 24' = 147 \text{ m}$
Vertical distance $AC = 147 \tan 5° \ 24' = -13.9 \text{ m}$
$\therefore$ Area of triangle $ABC = [S(S - a) \times (S - b) \times (S - c)]^{1/2}$

where $\quad S = \dfrac{1}{2}(157 + 246 + 147) = 275 \text{ m}$

$\therefore$ Area $= [275(275 - 157) \times (275 - 147) \times (275 - 246)]^{1/2}$

$\qquad = 10\,975 \text{ m}^2$

Assume level 0f $A = 100 \text{ m}$
then level of $B = 100 + 1.4 + 32.8 - 2.73 = 131.47 \text{ m}$
then level of $C = 100 + 1.4 - 13.9 - 2.82 = 84.68 \text{ m}$

$\therefore$ Depth of fill at $A = 31.47 \text{ m}$
Depth of fill at $C = 46.79 \text{ m}$
Vol of fill $=$ plan area $\times$ mean height

$$= 10\,975 \times \dfrac{1}{3}(31.47 + 46.79) = 286\,300 \text{ m}^3$$

*Example 5.8.* In order to find the radius of an existing road curve, three suitable points $A$, $B$ and $C$ were selected on its centre-line. The instrument was set at $B$ and the following readings taken on $A$ and $C$, the telescope being horizontal and the staff vertical.

| Staff at | Horizontal bearing | Stadia readings (m) | | |
|----------|--------------------|---------------------|--------|--------|
| A | 0° 00′ | 1.617 | 1.209 | 0.801 |
| C | 195° 34′ | 2.412 | 1.926 | 1.440 |

If the instrument has a constant of 100 and 0, calculate the radius of the circular arc $A, B, C$. If the trunnion axis was 1.54 m above the road at $B$, find the gradients $AB$ and $BC$.    (LU)

*Note :* As the theodolite is a clockwise-graduated instrument the angle $ABC$ as shown in *Figure 5.23* equals 195° 34′.

The angular relationships shown in the figure are from the geometry of angles at the centre being *twice* those at the circumference. It is therefore required to find angles $BAC$ and $BCA$ ($\alpha$ and $\beta$). From the formula for horizontal sights: $D = K_1 S + K_2$

$AB = 81.6 \text{ m} \qquad$ and $\qquad BC = 97.2 \text{ m}$

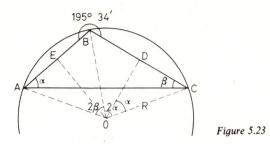

Figure 5.23

Assuming $AB$ is $0°$, then $BC = 15° 34'$ for 97.2 m

$\therefore$ Co-ords of $BC = 97.2 \dfrac{\sin}{\cos} 15° 34' = +26.08(\Delta E), +93.63(\Delta N)$

$\therefore$ Total co-ords of $C$ relative to $A = 26.08$ E; $(81.6 + 93.63)$ N $= 175.23$ N

Bearing $AC = \tan^{-1} \dfrac{26.08}{175.23} = 8° 28'$

$\therefore \alpha = 8° 28'$ and $\beta = (15° 34' - 8° 28') = 7° 06'$

In triangle $DCO$, $R = 48.6/\sin 8° 28' = 330$ m

Arc $AB = R \times 2\beta$ rad $= 330 \times 14° 12'$ rad $= 81.78$ m

Arc $BC = R \times 2\alpha$ rad $= 330 \times 16° 56'$ rad $= 97.53$ m

Grade $AB = (1.54 - 1.209)$ in $81.78 = 1$ in $250$ falling from $A$ to $B$
Grade $BC = (1.926 - 1.54)$ in $97.53 = 1$ in $250$ falling from $B$ to $C$

An alternative method of finding $\alpha$ and $\beta$ would have been to use the equation:

$$\tan \frac{A - C}{2} = \frac{a - c}{a + c} \tan \frac{A + C}{2}$$

However, using co-ordinates involves less computation and precludes the memorizing of equation in this case. This is particularly so in the next question where the above equation plus the sine rule would be necessary to find $CD$.

*Example 5.9.* The following readings were taken by a theodolite from station $B$ on to stations $A$, $C$ and $D$.

| Sight | Horizontal angle | Vertical angle | Stadia readings (m) | | |
| | | | Top | Centre | Bottom |
| --- | --- | --- | --- | --- | --- |
| A | 301°  10′ | | | | |
| C | 152°  36′ | −5°  00′ | 1.044 | 2.283 | 3.522 |
| D | 205°  06′ | +2°  30′ | 0.645 | 2.376 | 4.110 |

The line $BA$ in *Figure 5.24* has a bearing of $28° 46'$ and the instrument constants are 100 and 0. Find the slope and bearing of line $CD$.                    (LU)

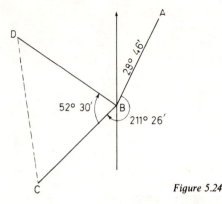

*Figure 5.24*

Distance $BC = 100S \cos^2 \theta = 247.8 \cos^2 5° = 246$ m
Height $BC \quad = 246 \tan 5° = -21.51$ m
Distance $BD = 346.5 \cos^2 2° 30' = 345.9$ m
Height $BD \quad = 345.9 \tan 2° 30' - 15.1$ m
Bearing $BC \quad = (28° 46' + 211° 26') = 240° 12'$
Bearing $BD \quad = (240° 12' + 52° 30') = 292° 42'$

$\therefore$  Co-ords of $BC = 246 \, \dfrac{\sin}{\cos} \, 240° 12' = -213.5 \, (\Delta E); \, -122.2 \, (\Delta N)$

Co-ords of $BD = 345.9 \, \dfrac{\sin}{\cos} \, 292° 42' = -319.2 \, (\Delta E); \, +133.5 \, (\Delta N)$

$\therefore$  Co-ords of $C$ relative to $D = -105.7 \, (\Delta E); \, +255.7 \, (\Delta N)$

$\therefore$  Bearing $CD = \tan^{-1} \dfrac{-105.7}{+255.7} = 327° 32'$

Length $CD = 255.7/\cos 22° 28' = 276.75$ m
Difference in level between $C$ and $D = -(21.51 + 2.283) - (15.1 - 2.376)$
$= 36.52$ m
$\therefore$  Grade $CD = 36.52$ in $276.75 = 1$ in $7.6$ rising

*Example 5.10.* Describe the essential features of a subtense bar and show how it is used in the determination of distance by a single measurement. Allowing for a 1″ error in the measurement of the angle, calculate from first principles the accuracy of the measurement of a distance of 60 m if a 2-m subtense bar is used. Show how the accuracy of such measurement varies with distance, and outline the method by which maximum accuracy will be obtained if subtense tacheometry is used in the determination of the distance between points situated on opposite banks of a river about 180 m wide.

(ICE)

Refer to *Section 5.3* for answer to first part of the question.

$$\text{Subtense angle } \theta'' = \frac{2 \times 206\,265}{60} = 6876''$$

$\therefore$  accuracy $= 1$ in $6876$

Refer equation (5.19) for variation of error with distance. Maximum accuracy may be obtained by the auxiliary base method. Refer *Section 5.3.3*.

*Example 5.11.* Specify briefly the apparatus required for subtense measurement giving a short description of the subtense bar. What accuracy could be expected in a length of a line measured in this way?

The following data refer to a subtense measurement between two stations:

Length of subtense bar (placed horizontally), 2 m with a standard error of 0.1 mm.
Angles subtended by ends of bar (from theodolite) $0° 32' 15''$ (10'') (12'') (18'') (16'') (14'') with a standard error of 4'' in each value.
Vertical circle readings, above horizontal $12° 10' 20''$ (18'') (25'') (21'') (21'') (17'') with a standard error of 4'' in each reading. The figures in brackets are the numbers of seconds obtained in five repeated readings, the degrees and minutes remaining unchanged. The standard error in the horizontality of the vertical circle index was 6''.

Determine: (1) The horizontal distance between the instrument axis and the mid-point of the subtense bar.
(2) The difference in level between these two points.
(3) The standard error in the level difference calculated in (2).     (LU)

(1) Mean horizontal angle $\theta = 0° 32' 14.2''$
  $\therefore D = b/2 \cot \theta/2 = \cot 0° 16' 07.1'' = 213.28$ m
(2) Mean vertical angle $= 12° 10' 20''$
  $\therefore H = D \tan \alpha = 213.28 \tan 12° 10' 20'' = 46.00$ m
(3) Standard error in length of bar $b = \delta b = \pm 0.1$ mm
  Standard error in angle $\theta = \delta\theta = \pm 4''/(6)^{1/2} = \pm 1.6''$
  Standard error in reading vertical angle $\alpha = \delta\alpha = $ as above $\pm 1.6''$

It is logical to assume that the vertical angles would be measured on each face of the instrument, thus cancelling out the vertical circle index error of 6''.

Now, differentiating with respect to each variable, the basic formula $H = D \tan \alpha$ gives

$$dH = \delta D \tan \alpha \quad \text{and} \quad \delta H = D \sec^2 \alpha\, \delta\alpha$$
$$\therefore \delta H = \pm[\tan^2 \alpha\, \delta D^2 + D^2 \sec^4 \alpha\, \delta\alpha^2]^{1/2}$$

but $\delta D$ is unknown and must first be found as follows:

$$D = b/\theta$$

$$\therefore \delta D = \frac{\delta b}{\theta} \quad \text{but} \quad \frac{1}{\theta} = \frac{D}{b} \quad \therefore \delta D = \frac{D\,\delta b}{b}$$

$$\delta D = \frac{-b}{\theta^2}\delta\theta \quad \text{but} \quad D^2 = \frac{b^2}{\theta^2} \quad \therefore \delta D = \frac{D^2\,\delta\theta}{b}$$

$$D = \pm\left[\frac{D^2\,\delta b^2}{b^2} + \frac{D^4\,\delta\theta^2}{b^2}\right]^{1/2} = \pm\frac{D}{b}[\delta b^2 = D^2\,\delta\theta^2]^{1/2}$$

$$\therefore \delta D = \pm\frac{213}{2}\left\{0.0001^2 + \frac{213^2 \times 1.6^2}{206\,265^2}\right\}^{1/2} = \pm 0.18 \text{ m}$$

$$\therefore \delta H = \pm[\tan^2 (12° 10' 20'') \times 0.18^2 + 213^2 \sec^4 (12° 10' 20'') \times 1.6^2]^{1/2}$$
$$= \pm 0.04 \text{ m}$$

## Student note

(1) All small angles in error computations ($\delta\theta$, $\delta\alpha$) must be changed to radians.
(2) The quantities involved need not be taken to any great accuracy as the relationship of the errors to the measured quantity is very small.
(3) For students unfamiliar with the theory of errors, the standard error in the angles is deduced as follows:

$\theta$ is the arithmetic mean of six measures

i.e.    $\theta = (\theta_1 + \theta_2 + \theta_3 + \theta_4 + \theta_5 + \theta_6)/6$

each measure is affected by an error $e$ of $\pm 4''$

$$\therefore \quad \theta \pm e = [(\theta_1 + e_1) + (\theta_2 + e_2) + (\theta_3 \pm e_3) + (\theta_4 \pm e_4) + (\theta_5 \pm e_5) + (\theta_6 \pm e_6)]/6$$

Subtracting    $\pm e = (\pm e_1 \pm e_2 \pm e_3 \pm e_4 \pm e_5 \pm e_6)/6$

assuming    $e_1 = e_2 = e_3$, etc. $= e_s$

$$\therefore \quad e^2 = \frac{6e_s^2}{6^2} \qquad \therefore \quad e = \pm \left(\frac{e_s^2}{6}\right)^{1/2} = \pm \frac{e_s}{(n)^{1/2}}$$

## Exercises

(*5.1*) In order to survey an existing road, three points $A$, $B$, and $C$ were selected on its centre-line. The instrument was set at $A$ and the following observations were taken.

| Staff | Horizontal angle | Vertical angle | Stadia readings (m) | | |
|-------|------------------|----------------|-------|-------|-------|
| B | 0° 00′ | −1° 11′ 20″ | 1.695 | 1.230 | 0.765 |
| C | 6° 29′ | −1° 04′ 20″ | 2.340 | 1.500 | 0.660 |

If the staff was vertical and the instrument constants 100 and 0, calculate the radius of the curve $ABC$. If the instrument was 1.353 m above $A$, find the falls $A$ to $B$ and $B$ to $C$.

(LU)

(*Answer:* $R = 337.8$ m, $A - B = 1.806$ m, $B - C = 1.482$ m)

(*5.2*) Readings were taken on a vertical staff held at points $A$, $B$ and $C$ with a tacheometer whose constants were 100 and 0. If the horizontal distances from instrument to staff were respectively 45.9, 63.6 and 89.4 m, and the vertical angles likewise $+5°$, $+6°$ and $-5°$, calculate the staff intercepts. If the mid-hair reading was 2.100 m in each case, what was the difference in level between $A$, $B$ and $C$?

(LU)

(*Answer:* $S_A = 0.462$, $S_B = 0.642$, $S_C = 0.900$, $B$ is 2.670 m above $A$, $C$ is 11.835 m below $A$)

(*5.3*) A theodolite has a multiplying constant of 100 and an additive constant of zero. When set 1.35 m above station $B$, the following readings were obtained.

| Stn | Sight | Horizontal circle | Vertical circle | Stadia readings (m) |
|-----|-------|-------------------|-----------------|---------------------|
| B | A | 28° 21′ 00″ | | |
| B | C | 82° 03′ 00″ | 20° 30′ | 1.140  2.292  3.420 |

The co-ordinates of A are E 163.86, N 0.0, and those of B, E 163.86, N 118.41. Find the co-ordinates of C and its height above datum if the level of B is 27.3 m AOD.     (LU)

(*Answer:* E 2.64 N 0.0, 101.15 m AOD)

(*5.4*) (a) The following observations were made to a 2-m subtense bar set 1.372 m above the ground:

Mean horizontal angle =     0° 20′ 30″
Mean vertical angle     = +05° 20′ 00″

Determine: (1) The horizontal distance between theodolite and bar.
            (2) The level of the subtense bar station, if the theodolite was set 1.524 m above the ground station whose level was 56.58 m OD.

(b) If the standard error in the measurement of the horizontal angle between the ends of the bar was ±1″, what is the fractional error in the above distance? Using the same equipment, into how many bays would you split the distance to increase the accuracy to 1 in 500?     (KP)

(*Answer:* D = 335.40 m, 88.04 m OAD, 1 in 1230, three bays)

# 6

# Curves

In the geometric design of motorways, railways, pipelines, etc., the design and setting out of curves is an important aspect of the engineer's work.

The initial design is usually based on a series of straight sections whose positions are defined largely by the topography of the area. The intersections of pairs of straights are then connected by horizontal curves (refer *Section 6.2*). In the vertical design, intersecting gradients are connected by curves in the vertical plane.

Curves can be listed under three main headings, as follows:

(1) Circular curves of constant radius.
(2) Transition curves of varying radius (spirals).
(3) Vertical curves of parabolic form.


## 6.1 CIRCULAR CURVES

Two straights, $D_1 T_1$ and $D_2 T_2$ in *Figure 6.1*, are connected by a circular curve of radius $R$:

(1) The straights when projected forward, meet at $I$: the *intersection point*.
(2) The angle $\Delta$ at $I$ is called the *angle of intersection* or the *deflection angle*, and equals the angle $T_1 0 T_2$ subtended at the centre of the curve 0.
(3) The angle $\phi$ at $I$ is called the *apex angle*, but is little used in curve computations.
(4) The curve commences from $T_1$ and ends at $T_2$; these points are called the *tangent points*.
(5) Distances $T_1 I$ and $T_2 I$ are the tangent lengths and are equal to $R \tan \Delta/2$.
(6) The length of curve $T_1 A T_2$ is obtained from:

Curve length = $R\Delta$ where $\Delta$ is expressed in radians, or

Curve length = $\dfrac{\Delta°.100}{D°}$ where *degree of curve* is used (see *Section 6.1.1*)

(7) Distance $T_1 T_2$ is called the *main chord (C)*, and from the *Figure*

$$\sin \frac{\Delta}{2} = \frac{T_1 B}{T_1 0} = \frac{\frac{1}{2} \text{ chord } (C)}{R} \qquad \therefore \ C = 2R \sin \frac{\Delta}{2}$$

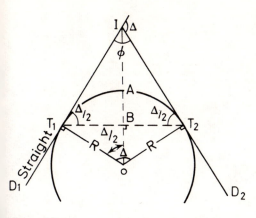

*Figure 6.1*

(8) *IA* is called the *apex distance* and equals

$$IO - R = R \sec \Delta/2 - R = R(\sec \Delta/2 - 1)$$

(9) *AB* is the *rise* and equals $R - OB = R - R \cos \Delta/2$

$$\therefore \quad AB = R(1 - \cos \Delta/2)$$

These equations should be deduced using a curve diagram (*Figure 6.1*) and not necessarily committed to memory.

## 6.1.1 Curve designation

Curves are designated either by their *radius* (*R*) or their *degree of curvature* (*D°*). The degree of curvature is defined as the angle subtended at the centre of a circle by an *arc* of 100 m (*Figure 6.2*).

Thus $\quad R = \dfrac{100 \text{ m}}{D \text{ rad}} = \dfrac{100 \times 180°}{D° \times \pi}$

$$\therefore \quad R = \frac{5729.578}{D°} \text{ m} \tag{6.1}$$

Thus a 10° curve has a radius of 572.9578 m.

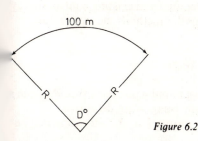

*Figure 6.2*

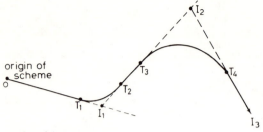

*Figure 6.3*

## 6.1.2 Through chainage

*Through chainage* is the horizontal distance from the start of a construction scheme. For instance, in *Figure 6.3* if the distance measured from 0 to $T_3$ is 2115.5 m, then it is said that the chainage of $T_3$ is 2115.5 m. If it was decided to set out the curve $T_3 T_4$ using 10-m chords, the first chord would be a sub-chord of 4.5 m. In this way the chainage at the end of the sub-chord would be in round figures, i.e. 2115.5 + 4.5 = 2120 m.

F requently questions supply the chainage at $I_2$ and require the chainage at $T_3$ and $T_4$, say. The chainage at $T_3$ is obtained by subtracting the tangent length $I_2 T_3$ from the chainage of $I_2$, whilst the chainage at $T_4$ is obtained by adding the curve length to the newly found chainage at $T_3$. This is logical as the curve is the route being constructed and the point $I_2$ is simply an established position to help curve setting out.

## 6.2 SETTING OUT CURVES

This is the process of establishing the centre-line of the curve on the ground by means of pegs at 10-m to 30-m intervals. In order to do this the tangent and intersection points must first be fixed in the ground, in their correct positions.

Consider *Figure 6.3*. The straights $0I_1, I_1 I_2, I_2 I_3$, etc., will have been designed on the plan in the first instance. Using railway curves, appropriate curves will now be designed to connect the straights. The tangent points of these curves will then be fixed making sure that the tangent lengths are equal, i.e. $T_1 I_1 = T_2 I_1$ and $T_3 I_2 = T_4 I_2$. The co-ordinates of the origin, point 0, and all the intersection points only will now be carefully scaled from the plan. Using these co-ordinates, the bearings of the straights are computed and, using the tangent lengths on these bearings, the co-ordinates of the tangent points are also computed. The difference of the bearings of the straights provides the deflection angles (Δ) of the curves which, combined with the tangent length, enables computation of the curve radius, through chainage and all setting-out data. Now the tangent and intersection points are set out from existing control survey stations and the curves ranged between them using the methods detailed below.

### 6.2.1 Setting out with theodolite and tape

The following method of setting out curves is the most popular and it is called *Rankine's deflection* or *tangential angle method*, the latter term being more definitive.

In *Figure 6.4* the curve is established by a series of chords $T_1 X, X Y$, etc. Thus, peg 1 a

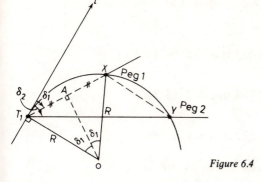

*Figure 6.4*

$X$ is fixed by sighting to $I$ with the theodolite reading zero, turning off the angle $\delta_1$ and measuring out the chord length $T_1X$ along this line. Setting the instrument to read the second deflection angle gives the direction $T_1Y$, and peg 2 is fixed by measuring the chord length $XY$ from $X$ until it intersects at $Y$. The procedure is now continued, the angles being set out from $T_1I$, and the chords measured from the previous station.

It is thus necessary to be able to calculate the setting out angles $\delta$ as follows:

Assume $0A$ bisects the chord $T_1X$ at right-angles, then

$$A\hat{T}_10 = 90° - \delta_1, \quad \text{but} \quad I\hat{T}_1 = 90°$$
$$\therefore I\hat{T}_1A = \delta_1$$

By radians arc length $T_1X = R2\delta_1$

$$\therefore \delta_1 \text{ rad} = \frac{\text{arc } T_1X}{2R} \approx \frac{\text{chord } T_1X}{2R}$$

$$\therefore \delta_1 \text{ min} = \frac{\text{chord } T_1X \times 180° \times 60}{2R.\pi} = 1718.9 \frac{\text{chord}}{R} \tag{6.2a}$$

or $\qquad \delta° = \dfrac{D° \times \text{chord}}{200} \qquad$ where *degree of curve* is used $\tag{6.2b}$

An example will now be worked to illustrate these principles.

The centre-line of two straights is projected forward to meet at $I$, the deflection angle being 30°. If the straights are to be connected by a circular curve of radius 200 m, tabulate all the setting-out data, assuming 20-m chords on a through chainage basis, the chainage of $I$ being 2259.59 m.

Tangent length $= R \tan \Delta/2 = 200 \tan 15° = 53.59$ m

$\quad \therefore$ Chainage of $T_1 = 2259.59 - 53.59 = 2206$ m
$\quad \therefore$ 1st sub-chord $= 14$ m

Length of circular arc $= R\Delta = 200(30°) \text{ rad} = 104.72$ m
From which the number of chords may now be deduced

$$\begin{array}{rl}
\text{i.e. 1st sub-chord} = & 14 \text{ m} \\
\text{2nd, 3rd, 4th, 5th chords} = & 20 \text{ m each} \\
\text{Final sub-chord} = & 10.72 \text{ m} \\
\text{Total} = & 104.72 \text{ m } (\textit{Check})
\end{array}$$

∴ Chainage of $T_2$ = 2206 m + 104.72 m = 2310.72 m

Deflection angles:

$$\text{For 1st sub-chord} = 1718.9 \frac{14}{200} = 120.3 \text{ min} = 2° \ 00' \ 19''$$

$$\text{Standard chord} = 1718.9 \frac{20}{200} = 171.9 \text{ min} = 2° \ 51' \ 53''$$

$$\text{Final sub-chord} = 1718.9 \frac{10.72}{200} = 92.1 \text{ min} = 1° \ 32' \ 08''$$

*Check:* The sum of the deflection angles = $\Delta/2$ = 14° 59′ 59″ ≈ 15°

| Chord number | Chord length (m) | Chainage (m) | Deflection angle ° ′ ″ | | | Setting-out angle ° ′ ″ | | | Remarks |
|---|---|---|---|---|---|---|---|---|---|
| 1 | 14 | 2220.00 | 2 | 00 | 19 | 2 | 00 | 19 | peg 1 |
| 2 | 20 | 2240.00 | 2 | 51 | 53 | 4 | 52 | 12 | peg 2 |
| 3 | 20 | 2260.00 | 2 | 51 | 53 | 7 | 44 | 05 | peg 3 |
| 4 | 20 | 2280.00 | 2 | 51 | 53 | 10 | 35 | 58 | peg 4 |
| 5 | 20 | 2300.00 | 2 | 51 | 53 | 13 | 27 | 51 | peg 5 |
| 6 | 10.72 | 2310.72 | 1 | 32 | 08 | 14 | 59 | 59 | peg 6 |

the error of 1″ is, in this case, due to the rounding-off of the angles to the nearest second and is negligible.

## 6.2.2 Setting out with two theodolites

Where chord taping is impossible, the curve may be set out using two theodolites at $T_1$ and $T_2$ respectively, the intersection of the lines of sight giving the position of the curve pegs.

The method is explained by reference to *Figure 6.5*. Set out the deflection angles from

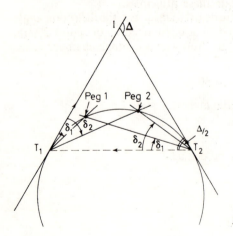

*Figure 6.5*

$T_1I$ in the usual way. From $T_2$, set out the same angles from the main chord $T_2T_1$. The intersection of the corresponding angles gives the peg position.

If $T_1$ cannot be seen from $T_2$, sight to $I$ and turn off the corresponding angles $\Delta/2 - \delta_1$, $\Delta/2 - \delta_2$, etc.

### 6.2.3 Setting out with two tapes (method of offsets)

Theoretically this method is exact, but in practice errors of measurement propagate round the curve. It is therefore generally used for minor curves.

In *Figure 6.6* line $OE$ bisects chord $T_1A$ at right-angles, then $ET_1O = 90° - \delta$, $\therefore CT_1A = \delta$, and triangles $CT_1A$ and $ET_1O$ are similar, thus

$$\frac{CA}{T_1A} = \frac{T_1E}{T_1O} \quad \therefore \quad CA = \frac{T_1E}{T_1O} \times T_1A$$

i.e. offset $\quad CA = \dfrac{\frac{1}{2}\text{chord} \times \text{chord}}{\text{radius}} = \dfrac{\text{chord}^2}{2R}$ (6.3)

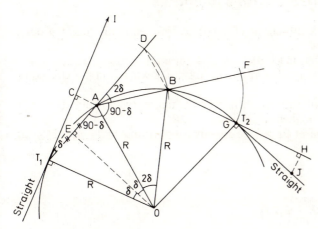

*Figure 6.6*

From *Figure 6.6*, assuming lengths $T_1A = AB = AD$

then angle $DAB = 2\delta$, and so offset $\quad DB = 2CA = \dfrac{\text{chord}^2}{R}$ (6.4)

The remaining offsets round the curve to $T_2$ are all equal to $DB$ whilst, if required, the offset $HJ$ to fix the line of the straight from $T_2$, equals $CA$.

The method of setting out is as follows:

It is sufficient to approximate distance $T_1C$ to the chord length $T_1A$ and measure this distance along the tangent to fix $C$. From $C$ a right-angled offset $CA$ fixes the first peg at $A$. Extend $T_1A$ to $D$ so that $AD$ equals a chord length; peg $B$ is then fixed by pulling out offset length from $D$ and chord length from $A$, and where they meet is the position $B$. This process is continued to $T_2$.

The above assumes equal chords. When the first or last chords are sub-chords, the following *Section 6.2.4* should be noted.

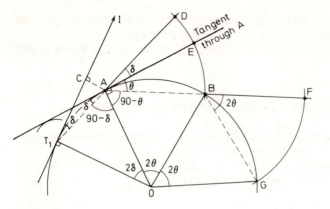

*Figure 6.7*

## 6.2.4 Setting out by offsets with sub-chords

In *Figure 6.7* assume $T_1A$ is a sub-chord of length $x$, from equation (6.3) the offset $CA = O_1 = x^2/2R$.

As the normal chord $AB$ differs in length from $T_1A$, the angle subtended at the centre will be $2\theta$ not $2\delta$. Thus, as shown in *Figure 6.6*, the offset $DB$ will not in this case equal $2CA$.

Construct a tangent through point $A$, then from the figure it is obvious that angle $EAB = \theta$, and if chord $AB = y$, then offset $EB = y^2/2R$.

Angle $DAE = \delta$, therefore offset $DE$ will be directly proportional to the chord length, thus:

$$DE = \frac{O_1}{x} \, y = \frac{x^2}{2R} \frac{y}{x} = \frac{xy}{2R}$$

Thus the total offset $DB = DE + EB$

$$= \frac{y}{2R}(x + y) \tag{6.5}$$

i.e.

$$= \frac{\text{Chord}}{2R}(\text{sub-chord} + \text{chord})$$

Thus having fixed $B$, the remaining offsets to $T_2$ are calculated as $y^2/R$ and set out in the usual way.

If the final chord is a sub-chord of length $x_1$, however, then the offset will be

$$\frac{x_1}{2R}(x_1 + y) \tag{6.6}$$

Students should note the difference between formulae (6.5) and (6.6).

A more practical approach to this problem, is actually to establish the tangent through $A$ in the field. This is done by swinging an arc of radius equal to $CA$, i.e. $x^2/2R$ from $T_1$. A line tangential to the arc and passing through peg $A$ will then be the required tangent from which offset $EB$, i.e. $y^2/2R$, may be set off.

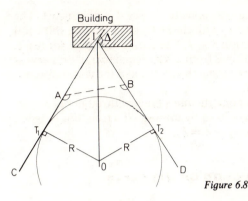

Figure 6.8

## 6.2.5 Setting out with inaccessible intersection point

In *Figure 6.8* it is required to fix $T_1$ and $T_2$, and obtain the angle $\Delta$, when $I$ is inaccessible.

Project the straights forward as far as possible and establish two points $A$ and $B$ on them. Measure distance $AB$ and angles $BAC$ and $DBA$ then:

angle $IAB = 180° - B\hat{A}C$ and angle $IBA = 180° - D\hat{B}A$, from which angle $BIA$ is deduced and angle $\Delta$. The triangle $AIB$ can now be solved for lengths $IA$ and $IB$. These lengths, when subtracted from the computed tangent lengths $(R \tan \Delta/2)$, give $AT_1$ and $BT_2$, which are set off along the straight to give positions $T_1$ and $T_2$ respectively.

## 6.2.6 Setting out with theodolite at an intermediate point on the curve

Due to an obstruction on the line of sight (*Figure 6.9*) or difficult communications and visibility on long curves, it may be necessary to continue the curve by ranging from a

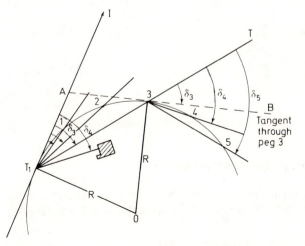

Figure 6.9

point on the curve. Assume that the setting-out angle to fix peg 4 is obstructed. The theodolite is moved to peg 3, backsighted to $T_1$ with the instrument reading 180°, and then turned to read 0°, thus giving the direction $3 - T$. The setting-out angle for peg 4, $\delta_4$, is turned off and the chord distance measured from 3. The remainder of the curve is now set off in the usual way, that is, $\delta_5$ is set on the theodolite and the chord distance measured from 4 to 5.

The proof of this method is easily seen by constructing a tangent through peg 3, then angle $A3T_1 = AT_13 = \delta_3 = T3B$. If peg 4 was fixed by turning off $\delta$ from this tangent, then the required angle from $3T$ would be $\delta_3 + \delta = \delta_4$.

### 6.2.7 Setting out with an obstruction on the curve

In this case (*Figure 6.10*) an obstruction on the curve prevents the chaining of the chord from 3 to 4. One may either

(1) Set out the curve from $T_2$ to the obstacle.
(2) Set out the chord length $T_14 = 2R \sin \delta_4$.

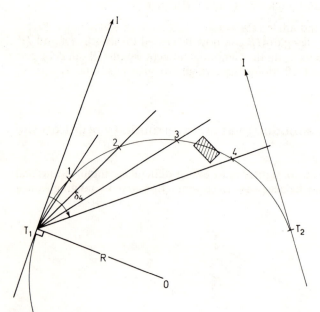

*Figure 6.10*

### 6.2.8 Passing a curve through a given point

In *Figure 6.11* it is required to find the radius of a curve which will pass through a point $P$, the position of which is defined by the distance $IP$ at an angle of $\phi$ to the tangent. Consider triangle $IPO$:

angle $\beta = 90° - \Delta/2 - \phi$ (right-angled triangle $IT_2O$)

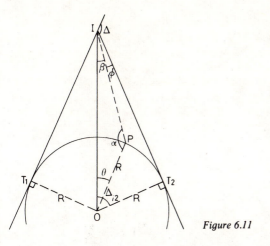

*Figure 6.11*

by sine rule:   $\sin \alpha = \dfrac{IO}{PO} \sin \beta$   but   $IO = R \sec \dfrac{\Delta}{2}$

$$\therefore \ \sin \alpha = \sin \beta \ \frac{R \sec \Delta/2}{R} = \sin \beta \sec \frac{\Delta}{2}$$

then   $\theta = 180° - \alpha - \beta$,   and by the sine rule:   $R = IP \dfrac{\sin \beta}{\sin \theta}$

## 6.3 COMPOUND AND REVERSE CURVES

Although equations are available which solve compound curves (*Figure 6.12*) and reverse curves (*Figure 6.13*), they are difficult to remember and students are advised to

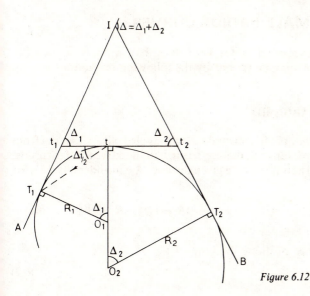

*Figure 6.12*

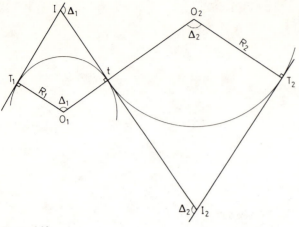

*Figure 6.13*

treat the problem as two simple curves with a common tangent point $t$.

In the case of the compound curve, the total tangent lengths $T_1 I$ and $T_2 I$ are found as follows:

$$R_1 \tan \Delta_1/2 = T_1 t_1 = t_1 t \quad \text{and} \quad R_2 \tan \Delta_2/2 = T_2 t_2 = t_2 t \text{ as } t_1 t_2 = t_1 t + t_2 t$$

then triangle $t_1 I t_2$ may be solved for lengths $t_1 I$ and $t_2 I$ which, if added to the known lengths $T_1 t_1$ and $T_2 t_2$ respectively, give the total tangent lengths.

In setting out this curve, the first curve $R_1$ is set out in the usual way to point $t$. The theodolite is moved to $t$ and backsighted to $T_1$, with the horizontal circle reading $(180° - \Delta_1/2)$. Set the instrument to read zero and it will then be pointing to $t_2$. Thus the instrument is now oriented and reading zero, prior to setting out curve $R_2$.

In the case of the reverse curve, both arcs can be set out from the common point $t$.

## 6.4 SHORT AND/OR SMALL-RADIUS CURVES

Short and/or small-radius curves such as for kerb lines, bay windows or for the construction of large templates, may be set out by the following methods.

### 6.4.1 Offsets from the tangent

The position of the curve (in *Figure 6.14*) is located by right-angled offsets $Y$ set out from distances $X$, measured along each tangent, thereby fixing half the curve from each side.

The offsets may be calculated as follows for a given distance $X$. Consider offset $Y_3$, for example.

$$\text{In} \quad \Delta ABO, \qquad AO^2 = OB^2 - AB^2 \qquad \therefore \quad (R - Y_3)^2 = R^2 - X_3^2 \qquad \text{and}$$
$$Y_3 = R - (R^2 - X_3^2)^{1/2}$$

thus for any offset $Y_i$ at distance $X_i$ along the tangent

$$Y_i = R - (R^2 - X_i^2)^{1/2} \tag{6.7}$$

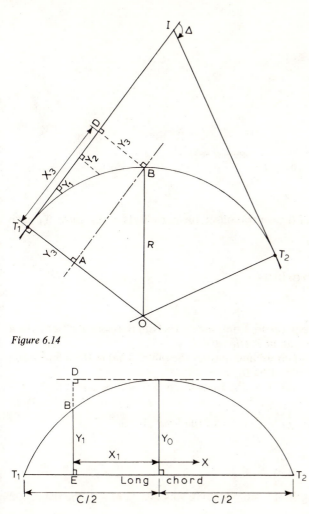

*Figure 6.14*

*Figure 6.15*

## 6.4.2 Offsets from the long chord

In this case (*Figure 6.15*) the right-angled offsets $Y$ are set off from the long chord $C$, at distances $X$ to each side of the centre offset $Y_0$.

An examination of *Figure 6.15* shows the central offset $Y_0$ equivalent to the distance $T_1A$ on *Figure 6.14*, thus:

$$Y_0 = R - [R^2 - (C/2)^2]^{1/2}$$

Similarly, $DB$ is equivalent to $DB$ on *Figure 6.14*, thus:    $DB = R - (R^2 - X_1^2)^{1/2}$

and offset $Y_1 = Y_0 - DB$    $\therefore$  $Y_1 = Y_0 - [R - (R^2 - X_1^2)^{1/2}]$

and for any offset $Y_i$ at distance $X_i$ each side of the

mid-point of $T_1T_2$:    $Y_i = Y_0 - [R - (R^2 - X^2)^{1/2}]$    (6.8)

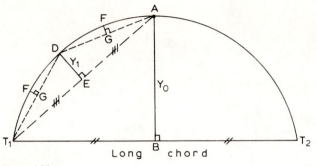

*Figure 6.16*

Therefore, after computation of the central offset, further offsets at distances $X_i$, each side of $Y_0$, can be found.

### 6.4.3 Halving and quartering

Referring to *Figure 6.16*:

(1) Join $T_1$ and $T_2$ to form the long chord. Compute and set out the central offset $Y_0$ to $A$ from $B$ (assume $Y_0 = 20$ m), as in *Section 6.4.2*.
(2) Join $T_1$ and $A$, now halve this chord and quarter the offset. That is, from mid-point $E$ set out offset $Y_1 = 20/4 = 5$ m to $D$.
(3) Repeat to give chords $T_1D$ and $DA$; the mid-offsets $FG$ will be equal to $Y_1/4 = 1.25$ m.

Repeat as often as necessary on both sides of the long chord.

### Worked examples

*Example 6.1.* The tangent length of a simple curve was 202.12 m and the deflection angle for a 30-m chord 2° 18′.
 Calculate the radius, the total deflection angle, the length of curve and the final deflection angle.                                                            (LU)

$$2° 18′ = 138′ = 1718.9 \frac{30}{R} \quad \therefore \ R = 373.67 \text{ m}$$

$$202.12 = R \tan \Delta/2 = 373.67 \tan \Delta/2 \quad \therefore \ \Delta = 56° 49′ 06″$$

Length of curve $= R\Delta$ rad $= 373.67 \times 0.991\ 667$ rad $= 370.56$ m
Using 30-m chords, the final sub-chord $= 10.56$ m

$$\therefore \ \text{final deflection angle} = \frac{138′ \times 10.56}{30} = 48.58′ = 0° 48′ 35″$$

*Example 6.2.* The straight lines *ABI* and *CDI* are tangents to a proposed circular curve of radius 1600 m. The lengths *AB* and *CD* are each 1200 m. The intersection point is

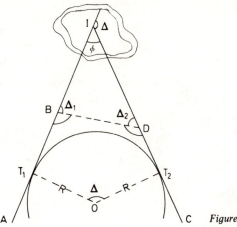

Figure 6.17

inaccessible so that it is not possible directly to measure the deflection angle; but the angles at $B$ and $D$ are measured as

$A\hat{B}D = 123° 48'$, $B\hat{D}C = 126° 12'$ and the length $BD$ is 1485 m

Calculate the distances from $A$ and $C$ of the tangent points on their respective straights and calculate the deflection angles for setting out 30-m chords from one of the tangent points.                                                          (LU)

Referring to *Figure 6.17*:

$\Delta_1 = 180° - 123° 48' = 56° 12'$,      $\Delta_2 = 180° - 126° 12' = 53° 48'$
$\therefore \Delta = \Delta_1 + \Delta_2 = 110°$
$\phi = 180° - \Delta = 70°$

Tangent lengths $IT_1$ and $IT_2 = R \tan \Delta/2 = 1600 \tan 55° = 2285$ m
By sine rule in triangle $BID$:

$$BI = \frac{BD \sin \Delta_2}{\sin \phi} = \frac{1485 \sin 53° 48'}{\sin 70°} = 1275.2 \text{ m}$$

$$ID = \frac{BD \sin \Delta_1}{\sin \phi} = \frac{1485 \sin 56° 15'}{\sin 70°} = 1314 \text{ m}$$

Thus     $AI = AB + BI = 1200 + 1275.2 = 2475.2$ m
         $CI = CD + ID = 1200 + 1314 = 2514$ m
$\therefore AT_1 = AI - IT_1 = 2475.2 - 2285 = 190.2$ m
         $CT_2 = CI - IT_2 = 2514 - 2285 = 229$ m

Deflection angle for 30-m chord $= 1718.9 \times 30/1600 = 32.23'$
$= 0° 32' 14''$

*Example 6.3.*   A circular curve of 800 m radius has been set out connecting two straights with a deflection angle of 42°. It is decided, for construction reasons, that the mid-point

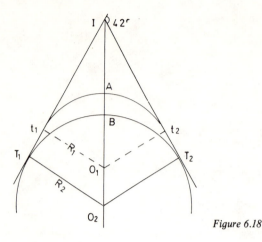

*Figure 6.18*

of the curve must be moved 4 m towards the centre, i.e. away from the intersection point. The alignment of the straights is to remain unaltered.

Calculate:

(1) The radius of the new curve.
(2) The distances from the intersection point to the new tangent points.
(3) The deflection angles required for setting out 30-m chords of the new curve.
(4) The length of the final sub-chord.                                    (LU)

Referring to *Figure 6.18*:

$$IA = R_1(\sec \Delta/2 - 1) = 800(\sec 21° - 1) = 56.92 \text{ m}$$
$$\therefore\ IB = IA + 4 \text{ m} = 60.92 \text{ m}$$

(1) Thus, $60.92 = R_2(\sec 21° - 1)$, from which $R_2 = 856$ m
(2) Tangent length $= IT_1 = R_2 \tan \Delta/2 = 856 \tan 21° = 328.6$ m

(3) Deflection angle for 30-m chord $= 1718.9\ C/R$ min $= 1718.9\ \dfrac{30}{856} = 1° 00' 14''$

(4) Curve length $= R\Delta$ rad $= \dfrac{856 \times 42° \times 3600}{206\ 265} = 627.5$ m

$\therefore$ Length of final sub-chord $= 27.5$ m

*Example 6.4.* The centre-line of a new railway is to be set out along a valley. The first straight *AI* bears 75°, whilst the connecting straight *IB* bears 120°. Due to site conditions it has been decided to join the straights with a compound curve.

The first curve of 500 m radius commences at $T_1$, situated 300 m from *I* on straight *AI*, and deflects through an angle of 25° before joining the second curve.

Calculate the radius of the second curve and the distance of the tangent point $T_2$ from *I* on the straight *IB*.                                    (KP)

Referring to *Figure 6.12*:

$$\Delta = 45°, \quad \Delta_1 = 25° \quad \therefore\ \Delta_2 = 20°$$

Tangent length $T_1 t_1 = R_1 \tan \Delta_1/2 = 500 \tan 12° 30' = 110.8$ m. In triangle $t_1 I t_2$

Angle $t_2 I t_1 = 180° - \Delta = 135°$
Length $I t_1 = T_1 I - T_1 t_1 = 300 - 110.8 = 189.2$ m

By sine rule

$$t_1 t_2 = \frac{I t_1 \sin t_2 I t_1}{\sin \Delta_2} = \frac{189.2 \sin 135°}{\sin 20°} = 391.2 \text{ m}$$

$$I t_2 = \frac{I t_1 \sin \Delta_1}{\sin \Delta_2} = \frac{189.2 \sin 25°}{\sin 20°} = 233.8 \text{ m}$$

$\therefore\ t t_2 = t_1 t_2 - T_1 t_1 = 391.2 - 110.8 = 280.4$ m
$\therefore\ 280.4 = R_2 \tan \Delta_2/2 = R_2 \tan 10°;\quad \therefore\ R_2 = 1590$ m

Distance $\ I T_2 = I t_2 + t_2 T_2 = 233.8 + 280.4$
$\qquad\qquad\qquad = 514.2$ m

*Example 6.5.* Two straights intersecting at a point $B$ have the following bearings, $BA$ 270°, $BC$ 110°. They are to be joined by a circular curve which must pass through a point $D$ which is 150 m from $B$ and the bearing of $BD$ is 260°.

Find the required radius, tangent lengths, length of curve and setting-out angle for a 30-m chord.   (LU)

Referring to *Figure 6.19*:
From the bearings, the apex angle $= (270° - 110°) = 160°$

$\therefore\ \Delta = 20°$

and angle $DBA = 10°$ (from bearings)

$\therefore\ OBD = \beta = 70°$

In triangle $BDO$ by sine rule

$$\sin \theta = \frac{OB}{OD} \sin \beta = \frac{R \sec \Delta/2}{R} \sin \beta = \sec \frac{\Delta}{2} \sin \beta$$

$\therefore\ \sin \theta = \sec 10° \times \sin 70°$
$\theta = \sin^{-1} 0.954\,190 = 72° 35' 25''\quad$ or
$(180° - 72° 35' 25'') = 107° 24' 35''$

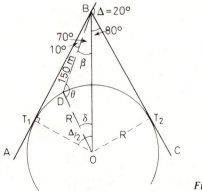

*Figure 6.19*

An examination of the figure shows that $\delta$ must be less than $10°$,

$$\therefore \quad \theta = 107° \ 24' \ 35''$$
$$\delta = 180° - (\theta + \beta) = 2° \ 35' \ 25''$$

By sine rule   $DO = R = \dfrac{DB \sin \beta}{\sin \delta} = \dfrac{150 \sin 70°}{\sin 2° \ 35' \ 25''}$

$$\therefore \quad R = 3119 \text{ m}$$

Tangent length $= R \tan \Delta/2 = 3119 \tan 10° = 550 \text{ m}$

Length of curve $= R\Delta$ rad $= \dfrac{3119 \times 20° \times 3600}{206 \ 265} = 1089 \text{ m}$

Deflection angle for 30-m chord $= 1718.9 \times \dfrac{30}{3119} = 0° \ 16' \ 32''$

*Example 6.6.* The co-ordinates in metres of two points $B$ and $C$ with respect to $A$ are:

B   470 E   500 N
C   770 E   550 N

Calculate the radius of a circular curve passing through the three points, and the co-ordinates of the intersection point $I$, assuming that $A$ and $C$ are tangent points.

(KP)

Referring to *Figure 6.20*:

*By co-ordinates*

Bearing   $AB = \tan^{-1} \dfrac{+470 \text{ E}}{+500 \text{ N}} = 43° \ 14'$

Bearing   $AC = \tan^{-1} \dfrac{+770 \text{ E}}{+550 \text{ N}} = 54° \ 28'$

Bearing   $BC = \tan^{-1} \dfrac{+300 \text{ E}}{+50 \text{ N}} = 80° \ 32'$

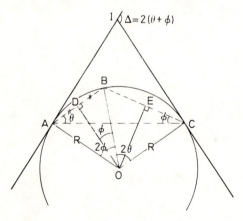

*Figure 6.20*

Distance   $AB = 500/\cos 43° 14' = 686$ m

From bearings of $AB$ and $AC$, angle $BAC = \theta = 11° 14'$
From bearings of $CA$ and $CB$, angle $BCA = \phi = 26° 04'$

as a check, the remaining angle, calculated from the bearings of $BA$ and $BC = 142° 42'$, summing to $180°$.

In right-angled triangle $DOB$:

$$OB = R = \frac{DB}{\sin \phi} = \frac{343}{\sin 26° 04'} = 781 \text{ m}$$

This result could now be checked through triangle $OEC$.

$$\Delta = 2(\phi + \theta) = 74° 36'$$

$$\therefore\ AI = R \tan \Delta/2 = 781 \tan 37° 18' = 595 \text{ m}$$

Bearing   $AI = $ bearing $AC - \Delta/2 = 54° 82' - 37° 18'$
$\qquad\qquad = 17° 10'$

$\therefore$ Co-ordinates of $I$ equal:

$$595 \, \frac{\sin}{\cos} \, 17° 10' = +176 \text{ E}, +569 \text{ N}$$

*Example 6.7.* Two straights $AEI$ and $CFI$, whose bearings are respectively $35°$ and $335°$, are connected by a straight from $E$ to $F$. The co-ordinates of $E$ and $F$ in metres are:

$E$    E 600.36   N 341.45
$F$    E 850.06   N 466.85

Calculate the radius of a connecting curve which shall be tangential to each of the lines $AE$, $EF$ and $CF$. Determine also the co-ordinates of $I$, $T_1$, and $T_2$, the intersection and tangent points respectively.                                             (KP)

Referring to *Figure 6.21*:

Bearing   $AI = 35°$,   bearing   $IC = (335° - 180°) = 155°$
$\qquad\qquad \therefore\ \Delta = 155° - 35° = 120°$

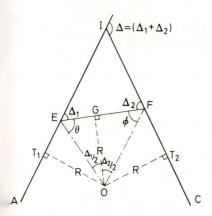

*Figure 6.21*

*By co-ordinates*

Bearing $EF = \tan^{-1} \dfrac{+249.70 \text{ E}}{+125.40 \text{ N}} = 63° \ 20'$

Length    $EF = 249.7/\sin 63° \ 20' = 279.42$ m

From bearings $AI$ and $EF$, angle $IEF = \Delta_1 = (63° \ 20' - 35°)$
$$= 28° \ 20'$$

From bearings $CI$ and $EF$, angle $IFE = \Delta_2 = (155° - 63° \ 20')$

$= 91° \ 40'$ check $(\Delta_1 + \Delta_2) = \Delta = 120°$

In triangle $EFO$

Angle $FEO = (90° - \Delta_1/2) = \theta = 75° \ 50'$
Angle $EFO = (90° - \Delta_2/2) = \phi = 44° \ 10'$
$$EG = GO \cot \theta = R \cot \theta$$
$$GF = GO \cot \phi = R \cot \phi$$

$\therefore \ EG + GF = EF = R(\cot \theta + \cot \phi)$

$\therefore \ R = \dfrac{EF}{(\cot \theta + \cot \phi)} = \dfrac{279.42}{\cot 75° \ 50' + \cot 44° \ 10'}$
$$= 217.97 \text{ m}$$
$$ET_1 = R \tan \Delta_1/2 = 217.97 \tan 14° \ 10' = 55.02 \text{ m}$$
$$FT_2 = R \tan \Delta_2/2 = 217.97 \tan 45° \ 50' = 224.4 \text{ m}$$
bearing $ET_1 = 215°$
bearing $FT_2 = 155°$

$\therefore$ Co-ordinates of $T_1 = 55.02 \ {}^{\sin}_{\cos} \ 215° = -31.56$ E, $-45.07$ N

$\therefore$ Total co-ordinates of $T_1 = $ E $600.36 - 31.56 = $ E $568.80$ m
$$= \text{N } 341.45 - 45.07 = \text{N } 296.38 \text{ m}$$

Similarly

Co-ordinates of $T_2 = 224.4 \ {}^{\sin}_{\cos} \ 155° = +98.84$ E, $-203.38$ N

$\therefore$ Total co-ordinates of $T_2 = $ E $850.06 + 94.84 = $ E $944.90$ m
$$- \text{N } 466.85 - 203.38 = \text{N } 263.47 \text{ m}$$

$T_1I = R \tan \Delta/2 = 217.97 \tan 60° = 377.54$ m

Bearing of $T_1I = 35°$

$\therefore$ Co-ordinates of $I = 377.54 \ {}^{\sin}_{\cos} \ 35° = +216.55$ E, $+309.26$ N

$\therefore$ Total co-ordinates of $I = $ E $586.20 + 216.55 = $ E $802.75$
$$= \text{N } 321.23 + 309.26 = \text{N } 630.49$$

The co-ordinates of $I$ can be checked via $T_2I$.

# Exercises

(*6.1*) In a town planning scheme, a road 9 m wide is to intersect another road 12 m wide at 60°, both being straight. The kerbs forming the acute angle are to be joined by a circular curve of 30 m radius and those forming the obtuse angle by one of 120 m radius.

Calculate the distances required for setting out the four tangent points.

Describe how to set out the larger curve by the deflection angle method and tabulate the angles for 15-m chords.    (LU)

(*Answer:* 75, 62, 72, 62 m. $\delta = 3°\ 35'$)

(*6.2*) A straight *BC* deflects 24° right from a straight *AB*. These are to be joined by a circular curve which passes through a point *P*, 200 m from *B* and 50 m from *AB*.

Calculate the tangent length, length of curve and deflection angle for a 30-m chord.    (LU)

(*Answer:* $R = 3754$ m, $IT = 798$ m, curve length $= 1572$ m, $0°\ 14'$)

(*6.3*) A reverse curve is to start at a point *A* and end at *C* with a change of curvature at *B*. The chord lengths *AB* and *BC* are respectively 661.54 m and 725.76 m and the radii likewise 1200 and 1500 m. Due to irregular ground the curves are to be set out using two theodolites and no tape or chain.

Calculate the data for setting out and describe the procedure in the field.    (LU)

(*Answer:* Tangent lengths: 344.09, 373.99; curve length: 670.2, 733, per 30-m chords: $\delta_1 = 0°\ 42'\ 54''$, $\delta_2 = 0°\ 34'\ 30''$)

(*6.4*) Two straights intersect making a deflection angle of 59° 24′, the chainage at the intersection point being 880 m. The straights are to be joined by a simple curve commencing from chainage 708 m.

If the curve is to be set out using 30-m chords on a through chainage basis, by the method of offsets from the chord produced, determine the first three offsets.

Find also the chainage of the second tangent point, and with the aid of sketches describe the method of setting out.    (KP)

(*Answer:* 0.066, 1.806, 2.985 m, 864.3 m)

(*6.5*) A circular curve of radius 250 m is to connect two straights, but in the initial setting out it soon becomes apparent that the intersection point is in an inaccessible position. Describe how it is possible in this case to determine by what angle one straight deflects from the other, and how the two tangent points may be accurately located and their through chainages calculated.

On the assumption that the chainages of the two tangent points are 502.2 m and 728.4 m, describe the procedure to be adopted in setting out the first three pegs on the curve by a theodolite (reading to 20″) and a steel tape from the first tangent point at 30-m intervals of through chainage, and show the necessary calculations.

If it is found to be impossible to set out any more pegs on the curve from the first tangent point because of an obstruction between it and the pegs, describe a procedure (without using the second tangent point) for accurately locating the fourth and succeeding pegs. No further calculations are required.    (ICE)

(*Answer:* 03° 11′ 10″, 06° 37′ 20″, 10° 03′ 40″)

## 6.5 TRANSITION CURVES

The *transition curve* is a curve of constantly changing radius. If used to connect a straight to a curve of radius $R$, then the commencing radius of the transition will be the same as the straight ($\infty$), and the final radius will be that of the curve $R$ (see *Figure 6.26*).

Consider a vehicle travelling at speed ($V$) along a straight. The forces acting on the vehicle will be its weight $W$, acting vertically down, and an equal and opposite force acting vertically up through the wheels. When the vehicle enters the curve of radius $R$ at tangent point $T_1$, an additional centrifugal force ($P$) acts on the vehicle, as shown in *Figures 6.22* and *6.23*. If $P$ is large the vehicle will be forced to the outside of the curve and may skid or overturn. In *Figure 6.23* the resultant of the two forces is shown as $N$,

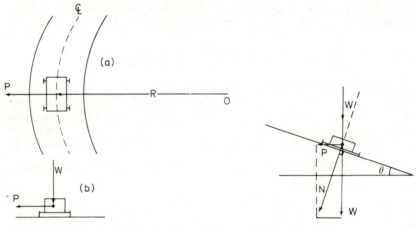

Figure 6.22                                Figure 6.23

and if the road is super-elevated normal to this force, there will be no tendency for the vehicle to skid. It should be noted that as

$$P = WV^2/Rg \qquad (6.9)$$

super-elevation will only cancel the effect of $P$ at a constant design speed $V$.

### 6.5.1  Principle of the transition

The purpose then of a transition curve is to:

(1) Achieve a gradual change of direction from the straight (radius $\infty$) to the curve (radius $R$).
(2) Permit the gradual application of super-elevation to counteract centrifugal force.

Since $P$ cannot be eliminated, it is allowed for by permitting it to increase uniformly along the curve. From equation (6.9), as $P$ is inversely proportional to $R$, the basic requirement of the ideal transition curve is that its radius should decrease uniformly with distance along it. This requirement also permits the uniform application of super-elevation, thus at distance $l$ along the transition the radius is $r$ and $rl = c$(constant)

$\therefore\ l/c = 1/r$

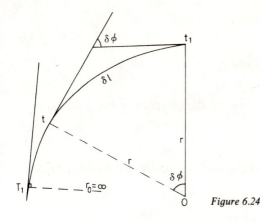

*Figure 6.24*

From *Figure 6.24*, $tt_1$ is an infinitely small portion of a transition $\delta l$ of radius $r$ thus:

$$\delta l = r\,\delta\phi$$

$\therefore\ 1/r = \delta\phi/\delta l$   which on substitution above gives

$$l/c = \delta\phi/\delta l$$

integrating:   $\phi = l^2/2c$   $\therefore\ l = (2c\phi)^{1/2}$

putting        $a = (2c)^{1/2}$

$$l = a(\phi)^{1/2} \tag{6.10}$$

when $c = RL$, $a = (2RL)^{1/2}$ and (6.10) may be written:

$$l = (2RL\phi)^{1/2} \tag{6.11}$$

the above expressions are for the *clothoid curve*, sometimes called the *Euler spiral*, which is the one most used in road design.

## 6.5.2 Curve design

The basic requirements in the design of transition curves are:

(1) The value of the minimum safe radius ($R$), and
(2) The length ($L$) of the transition curve. *Sections 6.5.5 or 6.5.6.*

The value $R$ may be found using either of the approaches *Sections 6.5.3 or 6.5.4.*

## 6.5.3 Centrifugal ratio

Centrifugal force is defined as $P = WV^2/Rg$; however, this 'overturning force' is counteracted by the mass ($W$) of the vehicle, and may be expressed as $P/W$, termed the *centrifugal ratio*. Thus, centrifugal ratio:

$$P/W = V^2/Rg \tag{6.12}$$

where $V$ is the design speed in m/s, $g$ is acceleration due to gravity in m/s² and $R$ is the minimum safe radius in metres.

When $V$ is expressed in km/h, the expression becomes

$$P/W = V^2/127R \tag{6.13}$$

Commonly-used values for centrifugal ratio are

0.21 to 0.25 on roads,     0.125 on railways

Thus, if a value of $P/W = 0.25$ is adopted for a design speed of $V = 50$ km/h then

$$R = \frac{50^2}{127 \times 0.25} = 79 \text{ m}$$

The minimum safe radius $R$ may be set either equal to or greater than this value.

### 6.5.4 Coefficient of friction

The alternative approach to find $R$ is based on Road Research Laboratory (RRL) values for the coefficient of friction between the car tyres and the road surface.

*Figure 6.25(a)* illustrates a vehicle passing around a correctly super-elevated curve. The resultant of the two forces is $N$. The force $F$ acting towards the centre of the curve is the friction applied by the car tyres to the road surface. These forces are shown in greater detail in *Figure 6.25(b)* from which it can be seen that

$$F_2 = \frac{WV^2}{Rg} \cos \theta, \quad \text{and} \quad F_1 = W \cos (90 - \theta) = W \sin \theta$$

$$\therefore \ F = F_2 - F_1 = \frac{WV^2}{Rg} \cos \theta - W \sin \theta$$

Similarly   $N_2 = \frac{WV^2}{Rg} \sin \theta, \quad \text{and} \quad N_1 = W \cos \theta$

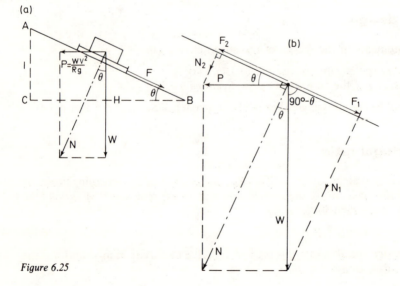

Figure 6.25

$$\therefore \quad N = N_2 + N_1 = \frac{WV^2}{Rg} \sin\theta + W\cos\theta$$

then
$$\frac{F}{N} = \frac{\dfrac{WV^2}{Rg}\cos\theta - W\sin\theta}{\dfrac{WV^2}{Rg}\sin\theta + W\cos\theta} = \frac{\dfrac{V^2}{Rg} - \tan\theta}{\dfrac{V^2}{Rg}\tan\theta + 1}$$

For Department of Transport (DoT) requirements, the maximum value for $\tan\theta = 1$ in $14.5 = 0.069$, and as $V^2/Rg$ cannot exceed 0.25 the term in the denominator can be ignored and

$$\frac{F}{N} = \frac{V^2}{Rg} - \tan\theta = \frac{V^2}{127R} - \tan\theta \qquad (6.14)$$

To prevent vehicles slipping sideways $F/N$ must be greater than the coefficient of friction $\mu$ between tyre and road. The RRL quote values for $\mu$ of 0.15, whilst 0.18 may be used up to 50 km/h, thus:

$$V^2/127R > \tan\theta + \mu \qquad (6.15)$$

For example, if the design speed is to be 100 km/h, super-elevation limited to 1 in 14.5 (0.069) and $\mu = 0.15$, then

$$\frac{100^2}{127R} = 0.069 + 0.15 \; .$$

$$\therefore \quad R = 360 \text{ m}$$

It can be seen that, provided that the super-elevation is always taken as 1 in 14.5, this approach is identical to the previous one.

### 6.5.5 Rate of application of super-elevation

It is recommended that on motorways super-elevation should be applied at a rate of 1 in 200, on all-purpose roads at 1 in 100, and on railways at 1 in 480. Thus, if on a motorway the super-elevation is computed as 0.5 m, then 100 m of transition curve would be required to accommodate 0.5 m at the required rate of 1 in 200, i.e. 0.5 in $100 = 1$ in 200. In this way the length $L$ of the transition is found.

The amount of super-elevation is obtained as follows:
From the triangle of forces in *Figure 6.25(a)*

$$\tan\theta = V^2/Rg = 1/H = 1 \text{ in } H$$

thus $H = Rg/V^2 = 127R/V^2$, where $V$ is in km/h

However, the DoT requirements for super-elevation are that it should

(1) Normally balance out 40% of the centrifugal force, thus 127 becomes 314 and

super-elevation $= 1$ in $314R/V^2$ \qquad (6.16)

(2) Not normally be greater than 1 in 14.5 (7%).
(3) Conform with the relationship expressed in formula (6.15).

Thus from formula (6.16) the rate of cross-fall is computed and, knowing the road width, the vertical amount of super-elevation is obtained. This amount, applied at a given rate, enables the length $L$ of the transition to be found.

### 6.5.6 Rate of change of radial acceleration

An alternative approach to finding the length of the transition is to use values for 'rate of change of radial acceleration' which would be unnoticeable to passengers when travelling by train. The appropriate values were obtained empirically by W. H. Shortt, an engineer working for the railways; hence it is usually referred to as *Shortt's Factor*.

Radial acceleration = $V^2/R$

Thus, as radial acceleration is inversely proportional to $R$ it will change at a rate proportional to the rate of change in $R$. The transition curve must therefore be long enough to ensure the rate of change of radius, and hence radial acceleration is unnoticeable to passengers.

Acceptable values for rate of change of radial acceleration ($q$) are 0.3 m/s³, 0.45 m/s³ and 0.6 m/s³.

Now, as radial acceleration is $V^2/R$ and the time taken to travel the length $L$ of the transition curve is $L/V$, then

$$\text{Rate of change of radial acceleration} = q = \frac{V^2}{R} \div \frac{L}{V} = \frac{V^3}{RL}$$

$$\therefore \ L = \frac{V^3}{Rq} = \frac{V^3}{3.6^3 . R . q} \tag{6.17}$$

where the design speed ($V$) is expressed in km/h

Although this method was originally devised for railway practice, it is also applied to road design.

## 6.6 SETTING-OUT DATA

*Figure 6.26* indicates the usual situation of two straights projected forward to intersect at $I$ with a clothoid transition curve commencing from tangent point $T_1$ and joining the circular arc at $t_1$. The second equal transition commences at $t_2$ and joins at $T_2$. Thus the composite curve from $T_1$ to $T_2$ consists of a circular arc with transitions at entry and exit.

### (1) Fixing the tangent points $T_1$ and $T_2$

In order to fix $T_1$ and $T_2$ the tangent lengths $T_1 I$ and $T_2 I$ are measured from $I$ back down the straights, or they are set out direct by co-ordinates.

$$T_1 I = T_2 I = (R + S) \tan \Delta/2 + C \tag{6.18}$$

where   $S = \text{shift} = L^2/24R - L^4/3! \times 7 \times 8 \times 2^3 R^3 + L^6/F! \times 11 \times 12$
$\times 2^5 R^5 - L^8/7! \times 15 \times 16 \times 2^7 R^7 \ldots$

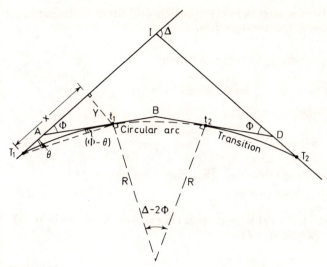

*Figure 6.26*

and   $C = L/2 - L^3/2! \times 5 \times 6 \times 2^2R^2 + L^5/4! \times 9 \times 10 \times 2^4R^4$
$\quad\quad - L^7/6! \times 13 \times 14 \times 2^6R^6 \ldots$

The values of $S$ and $C$ are abstracted from the *Highway Transition Curve Tables (Metric)* (see *Table 6.2*, p. 190).

## (2)  Setting out the transitions

Referring to *Figure 6.27*:

The theodolite is set at $T_1$ and oriented to $I$ with the horizontal circle reading zero. The transition is then pegged out using deflection angles ($\theta$) and chords (Rankine's method) in exactly the same way as for a simple curve.

The data are calculated as follows:

(a)  The length of transition $L$ is calculated (see design factors in *Sections 6.5.5* and *6.5.6*), assume $L = 100$ m.

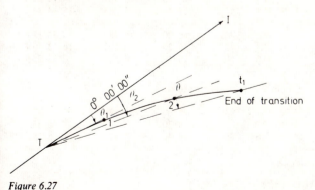

*Figure 6.27*

(b) It is then split into say 10 arcs, each 10 m in length (ignoring through chainage), the equivalent chord lengths being obtained from:

$$A - \frac{A^3}{24R^2} + \frac{A^5}{1920R^4}, \quad \text{where } A \text{ is the arc length}$$

(c) The setting out angles $\theta_1, \theta_2 \ldots \theta_n$ are obtained as follows:

Basic formula for clothoid: $l = (2RL\phi)^{1/2}$

$$\therefore \quad \Phi = \frac{l^2}{2RL} = \frac{L}{2R} \quad \text{when} \quad l = L \tag{6.19}$$

($l$ is any distance along the transition other than total distance $L$)

then   $\theta = \Phi/3 - 8\Phi^3/2835 - 32\Phi^5/467\,775 \ldots$ \hfill (6.20)

$= \Phi/3 - N$, where $N$ is taken from Tables and ranges in value from 0.1″ when $\Phi = 3°$, to 34′ 41.3″ when $\Phi = 86°$ (see *Table 6.1*).

Now   $\dfrac{\phi_1}{\Phi} = \dfrac{l_1^2}{L^2}$ \hfill (6.21)

$\therefore \quad \phi_1 = \Phi \dfrac{l_1^2}{L^2}$   where $l_1$ = chord length = 10 m, say

**TABLE 6.1 Interpolated deflection angles**

For any point on a spiral where angle consumed = $\phi$, the true deflection angle is $\phi/3$ minus the correction tabled below.
The back angle is $2\phi/3$ plus the same correction.

| Angle con- sumed $\phi$ | $\phi/3$ | Deduct N − | Deflection angle | Angle con- sumed $\phi$ | $\phi/3$ | Deduct N − | Deflection angle |
|---|---|---|---|---|---|---|---|
| ° | ° ′ | ′ ″ | ° ′ ″ | ° | ° ′ | ′ ″ | ° ′ ″ |
| 2 | 0 40 | NIL | 40 0 | 45 | 15 00 | 4 46.2 | 14 55 13.8 |
| 3 | 1 00 | 0.1 | 59 59.9 | 46 | 15 20 | 5 6.0 | 15 14 54.0 |
| 4 | 1 20 | 0.2 | 1 19 59.8 | 47 | 15 40 | 5 26.6 | 15 34 33.4 |
| 5 | 1 40 | 0.4 | 1 39 59.6 | 48 | 16 00 | 5 48.1 | 15 54 11.9 |
| 6 | 2 00 | 0.7 | 1 59 59.3 | 49 | 16 20 | 6 10.6 | 16 13 49.4 |
| 7 | 2 20 | 1.0 | 2 19 59.0 | 50 | 16 40 | 6 34.1 | 6 33 25.9 |
| | | | Continued at 1° intervals of $\phi$ | | | | |
| 41 | 13 40 | 3 35.9 | 13 36 24.1 | 84 | 28 00 | 32 14.4 | 27 27 45.6 |
| 42 | 14 00 | 3 52.1 | 13 56 7.7 | 85 | 28 20 | 33 26.9 | 27 46 33.1 |
| 43 | 14 20 | 4 9.4 | 14 15 50.6 | 86 | 28 40 | 34 41.3 | 28 5 18.7 |
| 44 | 14 40 | 4 27.4 | 14 35 32.6 | | | | |

*Reproduced by kind permission of the County Surveyors' Society*

and    $\theta_1 = \phi_1/3 - N_1$ (where $N_1$ is the value relative to $\phi$ in *Table 6.1*)

Similarly    $\phi_2 = \Phi \dfrac{l_2^2}{L^2}$    where    $l_2 = 20$ m

and    $\theta_2 = \phi_2/3 - N_2$    and so on.

Students should note:

(a) The values for $l_1$, $l_2$, etc. are accumulative.
(b) Thus the values obtained for $\theta_1$, $\theta_2$ etc. are the final setting-out angles and are obviously not to be summed.
(c) Although the chord length used is accumulative, the method of setting out is still the same as for the simple curve.

## (3)  Setting out circular arc $t_1t_2$

In order to set out the circular arc it is first necessary to establish the direction of the tangent $t_1B$ (*Figure 6.26*). The theodolite is set at $t_1$ and backsighted to $T_1$ with the horizontal circle reading $[180° - (\Phi - \theta)]$. Setting the instrument to zero will now orient it in the direction $t_1B$ with the circle reading zero, prior to setting out the simple circular arc. The angle $(\Phi - \theta)$ is called the *back-angle to the origin* and may be expressed as follows:

$$\theta = \Phi/3 - N$$
$$\therefore \ (\Phi - \theta) = \Phi - (\Phi/3 - N) = 2/3\Phi + N \qquad (6.22)$$

and is obtained direct from tables.

The remaining setting-out data are obtained as follows:

(a) As each transition absorbs an angle $\Phi$, then the angle subtending the circular arc $= (\Delta - 2\Phi)$.
(b) Length of circular arc $= R(\Delta - 2\Phi)$, which is then split into the required chord lengths $C$.
(c) The deflection angles $\delta$ min $= 1718.9 \ C/R$ are then set out from the tangent $t_1B$ in the usual way.

The second transition is best set out from $T_2$ to $t_2$. Setting out from $t_2$ to $T_2$ involves the 'osculating-circle' technique (refer *Section 6.9*).

The preceding formulae for clothoid transitions are specified in accordance with the latest *Highway Transition Curve Tables (Metric)* compiled by the County Surveyors' Society. As the equations involved in the setting-out data are complex, the information is generally taken straight from tables. However, approximation of the formulae produces two further transition curves, the *cubic spiral* and the *cubic parabola* (see *Section 6.7*).

In the case of the clothoid, *Figure 6.26* indicates an offset $Y$ at the end of the transition, distance $X$ along the straight, where

$$X = L - L^3/5 \times 4 \times 2! \, R^2 + L^5/9 \times 4^2 \times 4! \, R^4$$
$$- L^7/13 \times 4^3 \times 6! \, R^6 + \dots \qquad (6.23)$$
$$Y = L^2/3 \times 2R - L^4/7 \times 3! \times 2^3 R^3 \times L^6/11 \times 5! \times 2^5 R^5$$
$$- L^8/15 \times 7! \times 2^7 R^7 + \dots \qquad (6.24)$$

The clothoid is always set out by deflection angles, but the values for $X$ and $Y$ are useful in the large-scale plotting of such curves, and are taken from tables.

Refer Appendix B, p. 289, for derivation of clothoid formulae.

## 6.7 CUBIC SPIRAL AND CUBIC PARABOLA

Approximation of the clothoid formula produces the cubic spiral and cubic parabola, the latter being used on railway and tunnelling work because of the ease in setting out by offsets. The cubic spiral can be used for minor roads, as a guide for excavation prior to the clothoid being set out, or as a check on clothoid computation.

Approximating equation (6.24) gives

$Y = L^2/6R$,   which when   $L = l$,   $Y = y$,   becomes
$y = l^3/6RL$   (the equation for a cubic spiral)                        (6.25)

Approximating equation (6.23) gives

$X = L$,   thus   $x = l$

$\therefore \; y = x^3/6RL$   (the equation for a cubic parabola)                        (6.26)

In both cases:

Tangent length   $T_1I = (R + S)\tan \Delta/2 + C$
where                      $S = L^2/24R$                                        (6.27)
and                        $C = L/2$                                              (6.28)
                           $\Phi = L/2R = l^2/2RL$                                (6.29)
and                        $\theta = \Phi/3$                                      (6.30)

The deflection angles for these curves may be obtained as follows (the value of $N$ being ignored):

$\theta_1/\theta = l_1^2/L^2$   where $l$ is the chord/arc length                        (6.31)

When the value of $\Phi \approx 24°$, the radius of these curves starts to increase again, which makes them useless as transitions.

Refer to *Worked examples* (p. 191 etc.) for application of the above equations.

## 6.8 CURVE TRANSITIONAL THROUGHOUT

A curve transitional throughout (*Figure 6.28*) comprises two transitions meeting at a common tangent point $t$.

Tangent length   $T_1I = X + Y \tan \Phi$                        (6.32)

where $X$ and $Y$ are obtained from equations (6.23) and (6.24) and $\Phi = \Delta/2 = L/2R$.

$\therefore \; \Delta = L/R$                        (6.33)

## 6.9 THE OSCULATING CIRCLE

*Figure 6.29* illustrates a transition curve $T_1PE$. Through $P$, where the transition radius is $r$, a simple curve of the same radius is drawn and called the *osculating circle*.

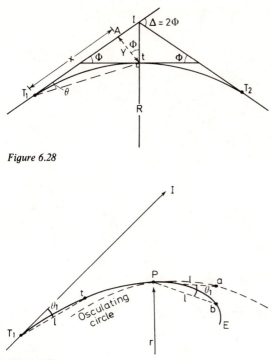

Figure 6.28

Figure 6.29

At $T_1$ the transition has the same radius as the straight $T_1I$, that is, $\infty$, but diverges from it at a constant rate. Exactly the same condition exists at $P$ with the osculating circle, that is, the transition has the same radius as the osculating circle, $r$, but diverges from it at a constant rate. Thus if chords $T_1t = Pa = Pb = l$, then

angle   $IT_1t = aPb = \theta_1$

This is the theory of the osculating circle, and its application is described in the following Sections.

## 6.9.1 Setting out with the theodolite at an intermediate point along the transition curve

*Figure 6.30* illustrates the situation where the transition has been set out from $T_1$ to $P_3$ in the normal way. The sight $T_1P_4$ is obstructed and the theodolite must be moved to $P_3$ where the remainder of the transition will be set out. The direction of the tangent $P_3E$ is first required from the back-angle $(\phi_3 - \theta_3)$.

From the *Figure* it can be seen that the angle from the tangent to the chord $P_3P_4'$ on the osculating circle is $\delta_1 = 1718.9l/r_3$ min. The angle between the chord on the osculating circle and that on the transition is $P_4'P_3P_4 = \theta_1$, thus the setting-out angle from the tangent to $P_4 = (\delta_1 + \theta_1)$, to $P_5 = (\delta_2 + \theta_2)$ and to $P_6 = (\delta_3 + \theta_3)$, etc.

For example, assuming $\Delta = 60°$, $L = 60$ m, $l =$ chord $= 10$ m, $R = 100$ m and $T_1P_3 = 30$ m, calculate the setting-out angles for the remainder of the transition from $P_3$.

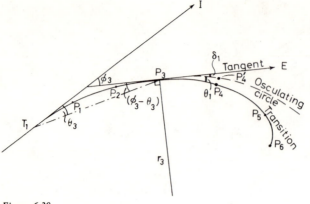

*Figure 6.30*

From basic formula:

$$\phi_3 = \frac{l_3^2}{2RL} = \frac{30^2}{2 \times 100 \times 60} = 4° \, 17' \, 50'' \; (-N_3, \text{ if clothoid})$$

or, if curve is defined by its 'degree of curvature' = $D$, then

$$\phi_3 = \frac{l_3^2}{200} \cdot \frac{D}{L} \; (-N_3, \text{ if clothoid})$$

thus the back-angle to the origin is found $(2/3)\phi_3$, and the tangent established as already shown.

Now from $\Phi = L/2R$ and $\theta = \Phi/3$, the angles $\theta_1$, $\theta_2$, and $\theta_3$ are found as normal. In practice these angles would already be available having been used to set out the first 30 m of the transition.

Before the angles to the osculating circle can be found the value of $r_3$ must be known, thus, from $rl = RL$

$$r_3 = RL/l_3 = 100 \times 60/30 = 200 \text{ m}$$

or 'degree of curvature' at $P_3$ 30 m from $T_1 = \dfrac{D}{L} \times l_3$

$$\therefore \; \delta_1 = 1718.9 \left( \frac{10}{200} \right) = 85.945 \text{ min} = 1° \, 25' \, 57''$$

$$\delta_2 = 2\delta_1 \quad \text{and} \quad \delta_3 = 3\delta_1 \quad \text{as for a simple curve}$$

The setting-out angles are then $(\delta_1 + \theta_1)$, $(\delta_2 + \theta_2)$, $(\delta_3 + \theta_3)$.

## 6.9.2 Setting out transition from the circular arc

*Figure 6.31* indicates the second transition in *Figure 6.26* to be set out from $t_2$ to $T_2$. The tangent $t_2D$ would be established by backsighting to $t_1$ with the instrument reading $[180° - (\Delta - 2\Phi)/2]$, setting to zero to fix direction $r_2D$. It can now be seen that the setting-out angles here would be $(\delta_1 - \theta_1)$, $(\delta_2 - \theta_2)$, etc. computed in the usual way.

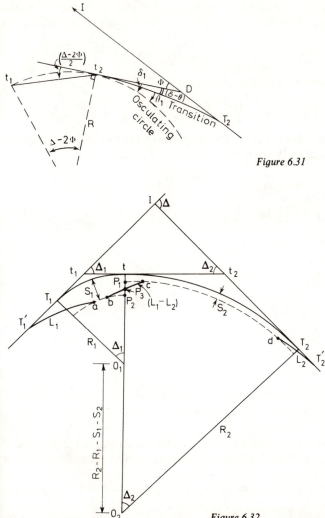

*Figure 6.31*

*Figure 6.32*

## 6.9.3 Transitions joining arcs of different radii (compound curves)

*Figure 6.12* indicates a compound curve requiring transitions at $T_1$, $t$ and $T_2$. To permit the entry of the transitions the circular arcs must be shifted forward as indicated in *Figure 6.32* where

$$S_1 = L_1^2/24R_1 \quad \text{and} \quad S_2 = L_2^2/24R_2$$

The lengths of transition at entry $(L_1)$ and exit $(L_2)$ are found in the normal way, whilst the transition connecting the compound arcs is

$$bc = L = (L_1 - L_2)$$

The distance $P_1P_2 = (S_1 - S_2)$ is bisected by the transition curve at $P_3$. The curve itself is bisected and length $bP_3 = P_3c$. As the curves at entry and exit are set out in the normal way, only the fixing of their tangent points $T_1'$ and $T_2'$ will be considered. In triangle $t_1It_2$

$$t_1t_2 = t_1t + tt_2 = (R_1 + S_1) \tan \Delta_1/2 + (R_2 + S_2) \tan \Delta_2/2$$

from which the triangle may be solved for $t_1I$ and $t_2I$.

Tangent length     $T_1'I = T_1't_1 + t_1I = (R_1 + S_1) \tan \Delta_1/2 + L_1/2 + t_1I$
and                     $T_2'I = T_2't_2 + t_2I = (R_2 + S_2) \tan \Delta_2/2 + L_2/2 + t_2I$

The curve $bc$ is drawn enlarged in *Figure 6.33*, from which the method of setting out, using the osculating circle, may be seen.

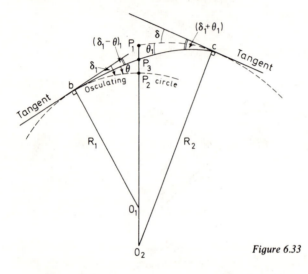

*Figure 6.33*

Setting out from $b$, the tangent is established from which the setting-out angles would be $(\delta_1 - \theta_1)$, $(\delta_2 - \theta_2)$, etc. as before, where $\delta_1$, the angle to the osculating circle, is calculated using $R_1$.

If setting out from $C$, the angles are obviously $(\delta_1 + \theta_1)$, etc., where $\delta_1$ is calculated using $R_2$.

Alternatively, the curve may be established by right-angled offsets from chords on the osculating circle, using the following equation:

$$y = \frac{x^3}{6RL} = \frac{x^3}{L^3} \frac{L^2}{6R} \quad \text{where} \quad \frac{L^2}{6R} = 4S$$

$$\therefore \quad y = \frac{4x^3}{L^3} (S_1 - S_2) \tag{6.34}$$

It should be noted that the osculating circle provides only an approximate solution, but as the transition is usually short, it may be satisfactory in practice. In the case of a reverse compound curve, *Figure 6.34*:

$$S = (S_1 + S_2), \quad L = (L_1 + L_2) \quad \text{and} \quad y = \frac{4x^3}{L^3} (S_1 + S_2)$$

otherwise it may be regarded as two separate curves.

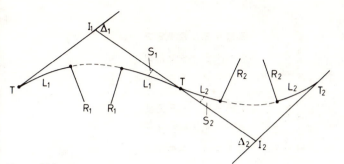

*Figure 6.34*

## 6.10 HIGHWAY TRANSITION CURVE TABLES (METRIC)

An examination of the complex equations defining the clothoid transition spiral indicates the obvious need for tables of prepared data to facilitate the design and setting out of such curves. These tables have been produced by the County Surveyors' Society under the title *Highway Transition Curve Tables (Metric)*, and contain a great deal of valuable information relating to the geometric design of highways. A very brief sample of the Tables is given here simply to convey some idea of the format and information contained therein.

As shown in *Section 6.6.3*, $\theta = \phi/3 - N$ and the 'back-angle' is $2\phi/3 + N$, all this information for various values of $\phi$ is supplied in *Table 6.1* and clearly shows that for large values of $\phi$, $N$ cannot be ignored.

Part only of *Table 6.2* is shown and it is the many tables like this that provide the bulk of the design data. Much of the information and its application to setting out should be easily understood by the student, so only a brief description of its use will be given here.

### Use of tables

(1) Check the angle of intersection of the straights ($\Delta$) by direct measurement in the field.
(2) Compare $\Delta$ with $2\Phi$, if $\Delta \leqslant 2\Phi$, then the curve is wholly transitional.
(3) Abstract $(R + S)$ and $C$ in order to calculate the tangent lengths $= (R + S) \tan \Delta/2 + C$.
(4) Take $\Phi$ from tables and calculate length of circular arc using $R(\Delta - 2\Phi)$, or, if working in 'degree of curvature' $D$, use

$$\frac{100(\Delta - 2\Phi)}{D}$$

(5) Derive chainages at the beginning and end of both transitions.
(6) Compute the setting-out angles for the transition $\theta_1 \ldots \theta$ from $\phi_1/\Phi = l_1^2/L^2$ from which $\theta_1 = \phi_1/3 - N_1$, and so on, for accumulative values of $l$.
(7) As control for the setting out, the end point of the transition can be fixed first by turning off from $T_1$ (the start of the transition) the 'deflection angle from the origin' $\theta$ and laying out the 'long chord' as given in the Tables. Alternatively, the right-angled offset $Y$ distance $X$ along the tangent may be used.

**TABLE 6.2**

Gain of accn, m/s³: 0.30, 0.45, 0.60
Speed value, km/h: 84.4, 96.6, 106.3
Increase in degree of curve per metre = $D/L$ = 0° 8' 0"
RL constant = 42 971.835
Degree of curvature based on 100 m standard arc

| Radius R (m) | Degree of curve D (° ' ") | | | Spiral length L (m) | Angle consumed φ (° ' ") | | | Shift S (m) | R + S (m) | C (m) | Long chord (m) | Co-ordinates X (m) | Y (m) | Deflection angle from origin (° ' ") | | | Back-angle to origin (° ' ") | | |
|---|---|---|---|---|---|---|---|---|---|---|---|---|---|---|---|---|---|---|---|
| 8594.3669 | 0 | 40 | 0 | 5 | 0 | 1 | 0 | 0.0001 | 8594.3670 | 2.5000 | 5.0000 | 5.0000 | 0.0005 | 0 | 0 | 20 | 0 | 0 | 40 |
| 4297.1835 | 1 | 20 | 0 | 10 | 0 | 4 | 0 | 0.0010 | 4297.1844 | 5.0000 | 10.0000 | 10.0000 | 0.0039 | 0 | 1 | 20 | 0 | 2 | 40 |
| 2864.7890 | 2 | 0 | 0 | 15 | 0 | 9 | 0 | 0.0033 | 2864.7922 | 7.5000 | 15.0000 | 15.0000 | 0.0131 | 0 | 3 | 0 | 0 | 6 | 0 |
| 2148.5917 | 2 | 40 | 0 | 20 | 0 | 16 | 0 | 0.0078 | 2148.5995 | 10.0000 | 20.0000 | 20.0000 | 0.0310 | 0 | 5 | 20 | 0 | 10 | 40 |
| 1718.8374 | 3 | 20 | 0 | 25 | 0 | 25 | 0 | 0.0152 | 1718.8885 | 12.5000 | 24.9999 | 24.9999 | 0.0606 | 0 | 8 | 20 | 0 | 16 | 40 |
| 1432.3945 | 4 | 0 | 0 | 30 | 0 | 36 | 0 | 0.0262 | 1432.4207 | 14.9999 | 29.9999 | 29.9997 | 0.1047 | 0 | 12 | 0 | 0 | 24 | 0 |
| 1227.7667 | 4 | 40 | 0 | 35 | 0 | 49 | 0 | 0.0416 | 1227.8083 | 17.4999 | 34.9997 | 34.9993 | 0.1663 | 0 | 16 | 20 | 0 | 32 | 40 |
| 1074.2959 | 5 | 20 | 0 | 40 | 1 | 4 | 0 | 0.0621 | 1074.3579 | 19.9998 | 39.9994 | 39.9986 | 0.2482 | 0 | 21 | 20 | 0 | 42 | 40 |

Continued for increasing values of $D$ up to 36° 00' 00"

*Reproduced by kind permission of the County Surveyors' Society*

(8) When the first transition is set out, set up the theodolite at the end point and with the theodolite reading $(180° - (\frac{2}{3}\Phi + N))$, backsight to $T_1$. Turn the theodolite to read $0°$ when it will be pointing in a direction tangential to the start of the circular curve prior to its setting out. This process has already been described.

(9) As a check on the setting out of the circular curve take $(R + S)$ and $S$ from the Tables to calculate the apex distance $= (R + S)$ $(\sec \Delta/2 - 1) + S$, from the intersection point $I$ of the straights to the centre of the circular curve.

(10) The constants $RL$ and $D/L$ are given at the head of the Tables and can be used as follows:

(a) Radius at any point $P$ on the transition $= r_p = RL/l_p$
(b) Degree of curve at $P = D_p = (D/L) \times l_p$, where $l_p$ is the distance to $P$ measured along the curve from $T_1$.

Similarly:

(c) Angle consumed at $P = \phi_p = \dfrac{l_p^2}{2RL}$ or $\dfrac{l_p^2}{200} \times \dfrac{D}{L}$

(d) Setting out angle from $T_1$ to $P = \theta_p = \dfrac{\phi_p}{3} - N_p$, or

$$\frac{l^2}{600} \times \frac{D}{L} - N_p$$

## Worked examples

*Example 6.8.* Part of a motorway scheme involves the design and setting out of a simple curve with cubic spiral transitions at each end. The transitions are to be designed such that the centrifugal ratio is 0.197, whilst the rate of change of centripetal acceleration is 0.45 m/s³ at a design speed of 100 km/h.

If the chainage of the intersection of the straights is 2154.22 m and the angle of deflection 50°, calculate:

(a) The length of transition to the nearest 10 m.
(b) The chainage at the beginning and the end of the total composite curve.
(c) The setting-out angles for the first three 10-m chords on a through chainage basis.

Briefly state where and how you would orient the theodolite in order to set out the circular arc. (KP)

Referring to *Figure 6.26*:

Centrifugal ratio $\quad P/W = V^2/127R$ $\qquad\qquad\qquad\qquad\qquad$ (6.13)

$$\therefore R = \frac{100^2}{127 \times 0.197} = 400 \text{ m}$$

Rate of change of centripetal acceleration $= q = \dfrac{V^3}{3.6^3 RL}$ $\qquad\qquad$ (6.17)

(a) $\quad \therefore L = \dfrac{100^3}{3.6^3 \times 400 \times 0.45} = 120 \text{ m}$

(b) To calculate chainage:

$$S = \frac{L^2}{24R} = \frac{120^2}{24 \times 400} = 1.5 \text{ m} \tag{6.27}$$

Tangent length $= (R + S) \tan \Delta/2 + L/2$ \hfill (6.18)

$$= (400 + 1.5) \tan 25° + 60 = 247.22 \text{ m}$$

$\therefore$ Chainage at $T_1 = 2154.22 - 247.22 = 1907$ m

To find length of circular arc:

length of circular arc $= R(\Delta - 2\Phi)$   where $\Phi = L/2R$

thus $\qquad\qquad 2\Phi = \dfrac{L}{R} = \dfrac{120}{400} = 0.3$ rad

and $\qquad\qquad\qquad \Delta = 50° = 0.872\,665$ rad

$\qquad \therefore\ R(\Delta - 2\Phi) = 400(0.872\,665 - 0.3) = 229.07$ m

Chainage at $\qquad T_2 = 1907.00 + 2 \times 120 + 229.07 = 2376.07$ m

(c) To find setting-out angles from equation (6.31) $\theta_1/\theta = l_1^2/L^2$

$$\theta = \frac{\Phi}{3} = \frac{L}{6R} = \frac{120}{6 \times 400} \text{ rad}$$

$$\theta'' = \frac{120 \times 206\,265}{6 \times 400} = 10\,313''$$

as the chainage of $T_1 = 1907$, then the first chord will be 3 m long to give a round chainage of 1910 m.

$$\therefore\ \theta_1 = \theta\frac{l_1^2}{L^2} = 10\,313'' \times \frac{3^2}{120^2} = 0° \ 00' \ 06.5''$$

$$\theta_2 = 10\,313'' \times \frac{13^2}{120^2} = 0° \ 02' \ 01''$$

$$\theta_3 = 10\,313'' \times \frac{23^2}{120^2} = 0° \ 06' \ 19''$$

For final part of answer refer to *Section 6.6(3)*.

*Example 6.9.* A transition curve of the cubic parabola type is to be set out from a straight centre-line. It must pass through a point which is 6 m away from the straight, measured at right-angles from a point on the straight produced, 60 m from the start of the curve.
    Tabulate the data for setting out a 120-m length of curve at 15-m intervals.
    Calculate the rate of change of radial acceleration for a speed of 50 km/h.

(LU)

    The above question may be read to assume that 120 m is only a part of the total transition length, thus $L$ is unknown.

From expression for a cubic parabola: $y = x^3/6RL = cx^3$

$$y = 6 \text{ m}, \quad x = 60 \text{ m} \quad \therefore \; c = \frac{1}{36\,000} = \frac{1}{6RL}$$

*The offsets are now calculated using this constant*

$$y_1 = \frac{15^3}{36\,000} = 0.094 \text{ m}$$

$$y_2 = \frac{30^3}{36\,000} = 0.750 \text{ m}$$

$$y_3 = \frac{45^3}{36\,000} = 2.531 \text{ m} \quad \text{and so on}$$

Rate of change of radial acceleration $= q = V^3/3.6^3 RL$

now $\quad \dfrac{1}{6RL} = \dfrac{1}{36\,000} \quad \therefore \; \dfrac{1}{RL} = \dfrac{1}{6000}$

$$\therefore \; q = \frac{50^3}{3.6^3 \times 6000} = 0.45 \text{ m/s}^3$$

*Example 6.10.* Two straights of a railway track of gauge 1.435 m have a deflection angle of 24° to the right. The straights are to be joined by a circular curve having cubic parabola transition spirals at entry and exit. The ratio of super-elevation to track gauge is not to exceed 1 in 12 on the combined curve, and the rate of increase/decrease of super-elevation on the spirals is not to exceed 1 cm in 6 m. If the through chainage of the intersection point of the two straights is 1488.8 m and the maximum allowable speed on the combined curve is to be 80 km/h, determine

(a) The chainages of the four tangent points.
(b) The necessary deflection angles (to the nearest 20″) for setting out the first four pegs past the first tangent point, given that pegs are to be set out at the 30-m points of the through chainage.
(c) The rate of change of radial acceleration on the curve when trains are travelling at the maximum permissible speed.　　　　　　　　　　　　　　　　(ICE)

(a) Referring to *Figure 6.26*, the four tangent points are $T_1, t_1, t_2, T_2$.

Referring to *Figure 6.25(a)*, as the super-elevation on railways is limited to 0.152 m, then $AB = 1.435 \text{ m} \approx CB$

$$\therefore \; \text{Super-elevation} = AC = \frac{1.435}{12} = 0.12 \text{ m} = 12 \text{ cm}$$

The rate of application $= 1$ cm in 6 m

$$\therefore \; \text{Length of transition} = L = 6 \times 12 = 72 \text{ m}$$

From *Section 6.5.5* $\quad \tan\theta = \dfrac{V^2}{127R} = \dfrac{1}{12}$

$$\therefore \; \frac{80^2}{127R} = \frac{1}{12} \quad \therefore \; R = 604.72 \text{ m}$$

$$\text{Shift} = S = \frac{L^2}{24R} = \frac{72^2}{24 \times 604.72} = 0.357 \text{ m}$$

Tangent length $= (R + S) \tan \Delta/2 + L/2 = 605.077 \tan 12° + 36$
$$= 164.6 \text{ m}$$

$\therefore$ Chainage $T_1 = 1488.8 - 164.6 = 1324.2$ m
Chainage $t_1 = 1324.2 + 72 = 1396.2$ m (end of transition)

To find length of circular curve:

$$2\Phi = \frac{L}{R} = \frac{72}{604.72} = 0.119\ 063 \text{ rad}$$

$\Delta = 24°$  $= 0.418\ 879$ rad

$\therefore$ Length of curve $= R(\Delta - 2\Phi) = 604.72(0.418\ 879 - 0.119\ 063)$
$$= 181.3 \text{ m}$$

$\therefore$ Chainage $t_2 = 1396.2 + 181.3 = 1577.5$ m

Chainage $T_2 = 1577.5 + 72$   $= 1649.5$ m

(b) From chainage of $T_1$ the first chord $= 5.8$ m

$$\theta = \frac{L}{6R} \times 206\ 265 = \frac{72 \times 206\ 265}{6 \times 604.72} = 4093''$$

$\therefore$ $\theta_1 = \theta \dfrac{l_1^2}{L^2} = 4093'' \times \dfrac{5.8^2}{72^2} = 27'' = 0° \ 00' \ 27''$    peg 1

$\theta_2 = 4093 \times \dfrac{35.8^2}{72^2} = 1012'' = 0° \ 16' \ 52''$    peg 2

$\theta_3 = 4093 \times \dfrac{65.8^2}{72^2} = 3418'' = 0° \ 56' \ 58''$    peg 3

$\theta_4 = 4093'' = \theta$ (end of transition) $= 1° \ 08' \ 10''$   peg 4

(c)    $q = \dfrac{V^3}{3.6^3 RL} = \dfrac{80^3}{3.6^3 \times 604.72 \times 72} = 0.25 \text{ m/s}^3$

*Example 6.11.* A compound curve *AB, BC* is to be replaced by a single arc with transition curves 100 m long at each end. The chord lengths *AB* and *BC* are respectively 661.54 and 725.76 m and radii 1200 m and 1500 m. Calculate the single arc radius:

(a) If *A* is used as the first tangent point.
(b) If *C* is used as the last tangent point.    (LU)

Referring to *Figure 6.12*, assume $T_1 = A$, $t = B$, $T_2 = C$, $R_1 = 1200$ m and $R_2 = 1500$ m. The requirements in this question are the tangent lengths *AI* and *CI*.

chord $AB = 2R_1 \sin \dfrac{\Delta_1}{2}$

$$\therefore \; \sin \frac{\Delta_1}{2} = \frac{661.54}{2 \times 1200}$$

$$\therefore \; \Delta_1 = 32°$$

Similarly,    $\sin \dfrac{\Delta_2}{2} = \dfrac{725.76}{3000}$

$$\therefore \; \Delta_2 = 28°$$

Distance    $At_1 = t_1 B = R_1 \tan \dfrac{\Delta_1}{2} = 1200 \tan 16° = 344 \text{ m}$

and    $Bt_2 = t_2 C = R_2 \tan \dfrac{\Delta_2}{2} = 1500 \tan 14° = 374 \text{ m}$

$$\therefore \; t_1 t_2 = 718 \text{ m}$$

By sine rule in triangle $t_1 I t_2$

$$t_1 I = \frac{718 \sin 28°}{\sin 120°} = 389 \text{ m}$$

and    $t_2 I = \dfrac{718 \sin 32°}{\sin 120°} = 439 \text{ m}$

$$\therefore \; AI = At_1 + t_1 I = 733 \text{ m}$$

$$CI = Ct_2 + t_2 I = 813 \text{ m}$$

To find single arc radius

(a) From tangent point $A$

$$AI = (R + S) \tan \Delta/2 + L/2$$

where    $S = L^2/24R$    and    $\Delta = \Delta_1 + \Delta_2 = 60°, \quad L = 100 \text{ m}$

then    $733 = \left( R + \dfrac{L^2}{24R} \right) \tan 30° + 50$    from which

$$R = 1182 \text{ m}$$

(b) From tangent point $C$

$$CI = (R + S) \tan \Delta/2 + L/2$$
$$813 = (R + L^2/24R) \tan 30° + 50 \quad \text{from which}$$
$$R = 1321 \text{ m}$$

*Example 6.12.* Two straights with a deviation angle of 32° are to be joined by two transition curves of the form $\lambda = a(\Phi)^{1/2}$ where $\lambda$ is the distance along the curve, $\Phi$ the angle made by the tangent with the original straight and $a$ as a constant.

The curves are to allow for a final 150 mm cant on a 1.435 m track, the straights being horizontal and the gradient from straight to full cant being 1 in 500.

Tabulate the data for setting out the curve at 15-m intervals if the ratio of chord to curve for 16° is 0.9872. Find the design speed for this curve.    (LU)

Referring to *Figure 6.28*:

Cant = 0.15 m, rate of application = 1 in 500,

therefore   $L = 500 \times 0.15 = 75$ m

As the curve is wholly transitional $\Phi = \Delta/2 = 16°$

$\therefore$ from   $\Phi = \dfrac{L}{2R}$,   $R = 134.3$ m

From ratio of chord to curve

Chord   $T_1 t = 75 \times 0.9872 = 74$ m
$\therefore$   $X = T_1 t \cos \theta = 73.7$ m   $(\theta = \Phi/3)$
$Y = T_1 t \sin \theta = 6.9$ m
$\therefore$   Tangent length $= X + Y \tan \Phi = 73.7 + 6.9 \tan 16° = 75.7$ m

Setting-out angles

$\theta_1 = (5° 20') \dfrac{15^2}{75^2} = 12' 48''$

$\theta_2 = (5° 20') \dfrac{30^2}{75^2} = 51' 12''$

and so on to $\theta_5$

Design speed

From *Figure 6.25(a)*   $\tan \theta \approx \dfrac{AC}{CB} = \dfrac{0.15}{1.435} = \dfrac{V^2}{Rg}$

from which   $V = 11.8$ m/s $= 42$ km/h

## Exercises

(6.6) The centre-line of a new road is being set out through a built-up area. The two straights of the road $T_1 I$ and $T_2 I$ meet giving a deflection angle of 45° and are to be joined by a circular arc with spiral transitions 100 m long at each end. The spiral from $T_1$ must pass between two buildings, the position of the pass point being 70 m along the spiral from $T_1$ and 1 m from the straight measured at right-angles.

Calculate all the necessary data for setting out the first spiral at 30-m intervals; thereafter find:

(a) The first three angles for setting out the circular arc, if it is to be set out by 10 equal chords.
(b) The design speed and rate of change of centripetal acceleration, given a centrifugal ratio of 0.1.
(c) The maximum super-elevation for a road width of 10 m.                    (KP)

(data $R = 572$ m, $T_1 I = 237.23$ m, $\theta_1 = 9' 01''$, $\theta_2 = 36' 37''$, $\theta_3 = 1° 40' 10''$)

(*Answer*: (a) 1° 44' 53", 3° 29' 46", 5° 14' 39", (b) 85 km/h, 0.23 m/s$^3$, (c) 1 m)

(6.7) A circular curve of 1800 m radius leaves a straight at through chainage 2468 m, joins a second circular curve of 1500 m radius at chainage 3976.5 m, and terminates on a second straight at chainage 4553 m. The compound curve is to be replaced by one of 2200 m radius with transition curves 100 m long at each end.

Calculate the chainages of the two new tangent points and the quarter point offsets of the transition curves. (LU)

(*Answer:* 2114.3 m, 4803.54 m; 0.012, 0.095, 0.32, 0.758 m)

(6.8) A circular curve must pass through a point *P* which is 70.23 m from *I*, the intersection point and on the bisector of the internal angle of the two straights *AI*, *IB*. Transition curves 200 m long are to be applied at each end and one of these must pass through a point whose co-ordinates are 167 m from the first tangent point along *AI* and 3.2 m at right angles from this straight. *IB* deflects 37° 54′ right from *AI* produced.

Calculate the radius and tabulate the data for setting out a complete curve.

(LU)

(*Answer:* $R = 1200$ m, $AI = IB = 512.5$ m, setting-out angles or offsets calculated in usual way)

(6.9) The limiting speed around a circular curve of 667 m radius calls for a super-elevation of 1/24 across the 10-m carriageway. Adopting the Department of Transport recommendations of a rate of 1 in 200 for the application of super-elevation along the transition curve leading from the straight to the circular curve, calculate the tangential angles for setting out the transition curve with pegs at 15-m intervals from the tangent point with the straight. (ICE)

(*Answer:* $L = 83$ m, 2′ 20″, 9′ 19″, 20′ 58″, 37′ 16″, 58′ 13″, 1° 11′ 18″)

(6.10) A circular curve of 610 m radius deflects through an angle of 40° 30′. This curve is to be replaced by one of smaller radius so as to admit transitions 107 m long at each end. The deviation of the new curve from the old at their mid-points is 0.46 m towards the intersection point.

Determine the amended radius assuming the shift can be calculated with sufficient accuracy on the old radius. Calculate the lengths of track to be lifted and of new track to be laid. (LU)

(*Answer:* $R = 590$ m, new track = 521 m, old track = 524 m)

(6.11) The curve connecting two straights is to be wholly transitional without intermediate circular arc, and the junction of the two transitions is to be 5 m from the intersection point of the straights which deflects through an angle of 18°.

Calculate the tangent distances and the minimum radius of curvature. If the super-elevation is limited to 1 vertical to 16 horizontal, determine the correct velocity for the curve and the rate of gain of radial acceleration. (LU)

(*Answer:* 95 m, 602 m, 68 km/h, 0.06 m/s³)

## 6.11 VERTICAL CURVES

Vertical curves (VC) are used to connect intersecting gradients in the vertical plane. Thus, in route design they are provided at all changes of gradient. They should be of

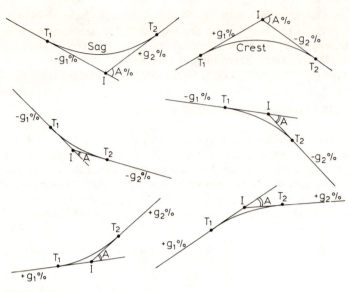

*Figure 6.35*

sufficiently large curvature to provide comfort to the driver, that is, they should have a low 'rate of change of grade'. In addition, they should afford adequate 'sight distances' for safe stopping at a given design speed.

The type of curve generally used to connect the intersecting gradients $g_1$ and $g_2$ is the simple parabola. Its use as a *sag* or *crest* curve is illustrated in *Figure 6.35*.

### 6.11.1 Gradients

In vertical-curve design the gradients are expressed as percentages, with a $-$ve for a downgrade and a $+$ve for an upgrade.

e.g.  A downgrade of 1 in 20 = 5 in 100 $= -5\% = -g_1\%$
An upgrade of 1 in 25 = 4 in 100 $= +4\% = +g_2\%$

The angle of deflection of the two intersecting gradients is called the *grade angle* and equals $A$ in *Figure 6.35*. The grade angle simply represents the change of grade through which the vertical curve deflects and is the algebraic difference of the two gradients

$$A\% = (g_1\% - g_2\%)$$

In the above example $A\% = (-5\% - 4\%) = -9\%$ ($-$ve indicates a sag curve).

### 6.11.2 Permissible approximations in vertical-curve computation

In civil engineering, road design is carried out in accordance with the following documents:

(1) Layouts of Roads in Rural Areas.
(2) Roads in Urban Areas.
(3) Motorway Design Memorandum.

However, practically all the geometric design in the above documents has been replaced by Department of Transport Standard TD 9/81, hereafter referred to simply as TD 9/81.

In TD 9/81 the desirable maximum gradients for vertical curve design are

| | |
|---|---|
| Motorways | 3% |
| Dual carriageways | 4% |
| Single carriageways | 6% |

Due to the shallowness of these gradients, the following VC approximations are permissible, thereby resulting in simplified computation (*Figure 6.36*).

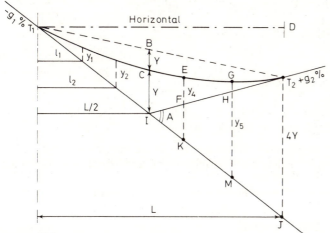

*Figure 6.36*

(1) Distance $T_1D = T_1BT_2 = T_1CT_2 = (T_1I + IT_2)$, without sensible error. This is very important and means that all distances may be regarded as horizontal in both the computation and setting out of vertical curves.
(2) The curve is of equal length each side of $I$. Thus $T_1C = CT_2 = T_1I = IT_2 = L/2$, without sensible error.
(3) The curve bisects $BI$ at $C$, thus $BC = CI = Y$ (the mid-offset).
(4) From similar triangles $T_1BI$ and $T_1T_2J$, if $BI = 2Y$, then $T_2J = 4Y$. $4Y$ represents the vertical divergence of the two gradients over half the curve length ($L/2$) and therefore equals $AL/200$.
(5) The basic equation for a simple parabola is

$$y = C.l^2$$

where $y$ is the *vertical* offset from gradient to curve, distance $l$ from the start of the curve and $C$ is a constant. Thus, as the offsets are proportional to distance squared, the following equation is used to compute them:

$$\frac{y_1}{Y} = \frac{l_1^2}{(L/2)^2} \tag{6.35}$$

where $Y =$ the mid-offset $= AL/800$ (refer 'VC computation', *Section 6.11.7*).

## 6.11.3 Vertical-curve design

In order to set out a vertical curve in the field, one requires levels along the curve at given chainage intervals. Before the levels can be computed, one must know the length $L$ of the curve. The value of $L$ is obtained from parameters supplied in *Table 3* of TD 9/81 (reproduced below as *Table 6.3*) and the appropriate parameters are $K$-values for specific design speeds and sight distances, then

$$L = KA \qquad (6.36a)$$

where $A$ = the difference between the two gradients (grade angle)

e.g. A $+4\%$ gradient is linked to a $-3\%$ gradient by a crest curve. What length of curve is required for a design speed of 100 km/h?

$$A = (4\% - (-3\%)) = +4\% \; (+\text{ve for crest})$$

From *Table 6.3*

C2 Desirable minimum crest $K$-value = 105
C3  Absolute minimum crest $K$-value =  59
   ∴ from   $L = K \cdot A$
Desirable minimum length = $L = 105 \times 7 = 735$ m
Absolute minimum length = $L =  59 \times 7 = 413$ m

Wherever possible the vertical and horizontal curves in the design process should be co-ordinated so that the sight distances are correlated and a more efficient overtaking provision is ensured.
The various design factors will now be dealt with in more detail.

**TABLE 6.3**

| Design speed (km/h) | 120 | 100 | 85 | 70 | 60 | 50 | $V^2/R$ |
|---|---|---|---|---|---|---|---|
| A *Stopping sight distance* (SSD), m | | | | | | | |
| A1 Desirable minimum | 300 | 225 | 165 | 125 | 95 | 70 | |
| A2 Absolute minimum | 225 | 165 | 125 | 95 | 70 | 50 | |
| B *Horizontal curvature*, m | | | | | | | |
| B1 Minimum $R^*$ without elimination of adverse camber and transitions | 2880 | 2040 | 1440 | 1020 | 720 | 510 | 5 |
| B2 Minimum $R^*$ with super-elevation of 2.5% | 2040 | 1440 | 1020 | 720 | 510 | 360 | 7.07 |
| B3 Minimum $R^*$ with super-elevation of 3.5% | 1440 | 1020 | 720 | 510 | 360 | 255 | 10 |
| B4 Desirable minimum $R$ super-elevation of 5% | 1020 | 720 | 510 | 360 | 255 | 180 | 14.14 |
| B5 Absolute minimum $R$ super-elevation of 7% | 720 | 510 | 360 | 255 | 180 | 127 | 20 |
| B6 Limiting radius super-elevation of 7% at sites of special difficulty (category B design speeds only) | 510 | 360 | 255 | 180 | 127 | 90 | 28.28 |
| C *Vertical curvature* | | | | | | | |
| C1 FOSD overtaking crest $K$-value | * | 400 | 285 | 200 | 142 | 100 | |
| C2 Desirable minimum* crest $K$-value | 185 | 105 | 59 | 33 | 19 | 11 | |
| C3 Absolute minimum crest $K$-value | 105 | 59 | 33 | 19 | 11 | 6.5 | |
| C4 Absolute minimum sag $K$-value | 37 | 26 | 20 | 20 | 20 | 13.5 | |
| D *Overtaking sight distance* | | | | | | | |
| D1 Full overtaking sight distance (FOSD), m | * | 580 | 490 | 410 | 345 | 290 | |

* Not recommended for use in the design of single carriageways (see Part C, Paras 2.7–2.8)
(*Reproduced with permission of the Controller of Her Majesty's Stationery Office*)

## (1) $K$-value

Rate of change of gradient ($r$) is the rate at which the curve passes from one gradient ($g_1\%$) to the next ($g_2\%$) and is similar in concept to rate of change of radial acceleration in horizontal transitions. When linked to design speed it is termed *rate of vertical acceleration* and should never exceed 0.3 m/sec$^2$.

A typical example of a badly-designed vertical curve with a high rate of change of grade, is a hump-backed bridge where usually the two approaching gradients are quite steep and connected by a very short length of vertical curve. Thus one passes through a large grade angle $A'$ in a very short time, with the result that often a vehicle will leave the ground and/or cause great discomfort to its passengers.

Commonly-used design values for $r$ are

3%/100 m on crest curves
1.5%/100 m on sag curves

thereby affording much larger curves to prevent rapid change of grade and provide adequate sight distances.

Working from first principles if $g_1 = -2\%$ and $g_2 = +4\%$ (sag curve), then the change of grade from $-2\%$ to $+4\% = 6\%$ ($A$), the grade angle. Thus, to provide for a rate of change of grade of 1.5%/100 m, one would require 400 m ($L$) of curve. If the curve was a crest curve, then using 3%/100 m gives 200 m ($L$) of curve

$$\therefore \ L = 100 \cdot A/r \tag{6.36b}$$

Now, expressing rate of change of grade as a single number we have

$$K = 100/r \tag{6.36c}$$

and as shown previously, $L = KA$.

## (2) Sight distances

*Sight distance* is a safety design factor which is intrinsically linked to rate of change of grade, and hence to $K$ values.

Consider once again the hump-backed bridge. Drivers approaching from each side of this particular vertical curve cannot see each other until they arrive, simultaneously, almost on the crest; by which time it may be too late to prevent an accident. Had the curve been longer and flatter, thus resulting in a low rate of change of grade, the drivers would have had a longer sight distance and consequently more time in which to take avoiding action.

Thus, sight distance, i.e. the length of road ahead that is visible to the driver, is a safety factor, and it is obvious that the sight distance must be greater than the stopping distance in which the vehicle can be brought to rest.

Stopping distance is dependent upon

(a) Speed of the vehicle.
(b) Braking efficiency.
(c) Gradient.
(d) Coefficient of friction between tyre and road.
(e) Road conditions.
(f) Driver's reaction time.

In order to cater for all the above variables, the height of the driver's eye above the road surface is taken as being only 1.05 m; a height applicable to sports cars whose braking efficiency is usually very high. Thus, other vehicles, such as lorries, with a much greater eye height would have a much longer sight distance in which to stop.

## (3)  Sight distances on crests

Sight distances on crests are defined as follows:

(a) *Stopping sight distance (SSD) (Figure 6.37(a))*
The SSD is measured from a driver's eye height of between 1.05 m and 2 m to an object of between 0.26 m and 2 m above the road surface. It should be checked in both the horizontal and vertical planes, between any two points in the centre of the lane, on the inside of the curve. Values per design speed are shown at A1 and A2 of *Table 6.3*. At least A2 should be provided on all single and dual carriageways.

(b) *Full overtaking sight distance (FOSD) (Figure 6.37(b))*
The above sight distance is for overtaking vehicles on single carriageways, since sufficient visibility for overtaking is required on as much of the road as possible. FOSD

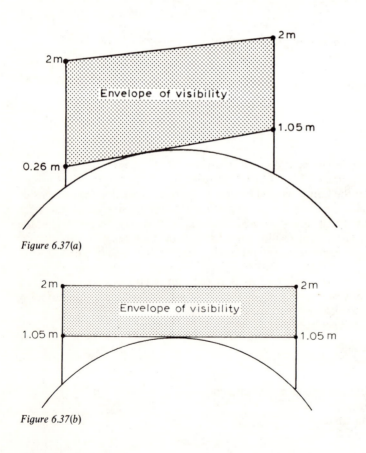

*Figure 6.37(a)*

*Figure 6.37(b)*

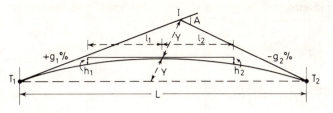

*Figure 6.38*

should also be checked in both the horizontal and vertical planes as specified for SSD. D1 in *Table 6.3* shows the appropriate distances for specific design speeds.

Although equations are unnecessary when using design tables, they can be developed to calculate curve lengths $L$ for given sight distances $S$, as follows:

(i)  When $S < L$ (*Figure 6.38*)

From basic equation $y = Cl^2$

$$Y = C(L/2)^2, \qquad h_1 = C(l_1)^2 \qquad \text{and} \qquad h_2 = C(l_2)^2$$

then
$$\frac{h_1}{Y} = \frac{l_1^2}{(L/2)^2} = \frac{4l_1^2}{L^2} \quad \text{and} \quad \frac{h_2}{Y} = \frac{4l_2^2}{L^2}$$

thus
$$l_1^2 = \frac{h_1 L^2}{4Y} \quad \text{but since} \quad 4Y = \frac{AL}{200}$$

$$l_1^2 = \frac{200 h_1 L}{A}$$

and
$$l_1 = (h_1)^{1/2}\left(\frac{200L}{A}\right)^{1/2}$$

Similarly
$$l_2 = (h_2)^{1/2}\left(\frac{200L}{A}\right)^{1/2}$$

$$\therefore \; S = (l_1 + l_2) = [(h_1)^{1/2} + (h_2)^{1/2}]\left(\frac{200L}{A}\right)^{1/2} \tag{6.37}$$

and
$$L = \frac{S^2 A}{200[(h_1)^{1/2} + (h_2)^{1/2}]^2} \tag{6.38}$$

when
$$h_1 = h_2 = h$$

$$L = \frac{S^2 A}{800h} \tag{6.39}$$

(ii)  When $S > L$ it can similarly be shown that

$$L = 2S - \frac{200}{A}[(h_1)^{1/2} + (h_2)^{1/2}]^2 \tag{6.40}$$

and when $h_1 = h_2 = h$

$$L = 2S - \frac{800h}{A} \tag{6.41}$$

When $S = L$, substituting in either of equations (6.39) or (6.41) will give the correct solution.

e.g. (6.39) $\quad L = \dfrac{S^2 A}{800h} = \dfrac{L^2 A}{800h} = \dfrac{800h}{A}$

and (6.41) $\quad L = 2S - \dfrac{800h}{A} = 2L - \dfrac{800h}{A} = \dfrac{800h}{A}$

N.B. If the relationship of $S$ to $L$ is not known then both cases must be considered; one of them will not fulfil the appropriate argument $S < L$ or $S > L$ and is therefore wrong.

## (4) Sight distances on sags

Visibility on sag curves is not obstructed as it is in the case of crests; thus sag curves are designed for at least absolute minimum comfort criteria of 0.3 m/sec². However, for design speeds of 70 km/h and below in unlit areas, sag curves are designed to ensure that headlamps illuminate the road surface for at least absolute minimum SSD. The relevant $K$ values are given in C4, *Table 6.3*.

The headlight is generally considered as being 0.75 m above the road surface with its beam tilted down at 1° to the horizontal. As in the case of crests, equations can be developed if required.

Consider *Figure 6.39 where L is greater than S*. From the equation for offsets:

$$\frac{BC}{T_2 D} = \frac{S^2}{L^2} \quad \therefore \; BC = \frac{S^2 (T_2 D)}{L^2}$$

but $T_2 D$ is the vertical divergence of the gradients and equals

$$\frac{A.L}{100.2} \quad \therefore \; BC = \frac{A.S^2}{200L} \qquad\qquad (a)$$

also $\qquad\qquad BC = h + S \tan x° \qquad\qquad\qquad\qquad (b)$

Equating $(a)$ and $(b)$: $\quad L = S^2 A (200h + 200S \tan x°)^{-1} \qquad (6.42)$

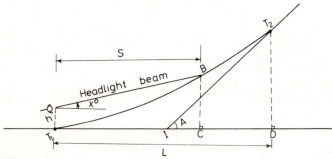

Figure 6.39

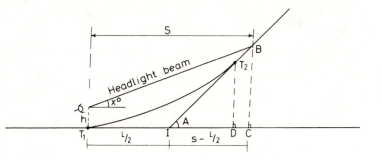

*Figure 6.40*

putting $x° = 1°$ and $h = 0.75$ m

$$L = \frac{S^2 A}{150 + 3.5S} \tag{6.43}$$

Similarly, *when S is greater than L* (*Figure 6.40*)

$$BC = \frac{A}{100}\left(S - \frac{L}{2}\right) = h + S \tan x°$$

equating:  $L = 2S - (200h + 200S \tan x°)/A$  (6.44)

when $x° = 1°$ and $h = 0.75$ m

$$L = 2S - (150 + 3.5S)/A \tag{6.45}$$

## 6.11.4 Passing a curve through a point of known level

In order to ensure sufficient clearance at a specific point along the curve it may be necessary to pass the curve through a point of known level. For example, if a bridge parapet or road furniture were likely to intrude into the envelope of visibility, it would be necessary to design the curve to prevent this.

This technique will be illustrated by the following example. A downgrade of 4% meets a rising grade of 5% in a sag curve. At the start of the curve the level is 123.06 m at chainage 3420 m, whilst at chainage 3620 m there is an overpass with an underside level of 127.06 m. If the designed curve is to afford a clearance of 5 m at this point, calculate the required length (*Figure 6.41*).

*To find the offset distance CE.*

From chainage horizontal distance $T_1 E = 200$ m at $-4\%$

∴  Level at $E = 123.06 - 8 = 115.06$ m

Level at $C = 127.06 - 5 = 122.06$ m

∴ Offset $CE = 7$ m

*From offset equation* $\dfrac{CE}{T_2 B} = \dfrac{(T_1 E)^2}{(T_1 B)^2}$

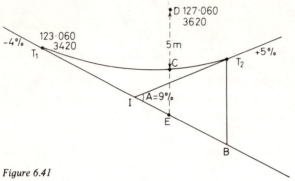

*Figure 6.41*

but $T_2B$ = the vertical divergence = $\dfrac{A}{100}\dfrac{L}{2}$, where $A = 9$

$$\therefore\ CE = \frac{AL}{200}\frac{200^2}{L^2} = \frac{1800}{L}$$

$$\therefore\ L = 257 \text{ m}$$

### 6.11.5 To find the chainage of highest or lowest point on the curve

The position and level of the highest or lowest point on the curve is frequently required for drainage design.

With reference to *Figure 6.41*, if one considers the curve as a series of straight lines, then at $T_1$ the grade of the line is $-4\%$ gradually changing throughout the length of the curve until at $T_2$ it is $+5\%$. There has thus been a change of grade of 9% in distance $L$. At the lowest point the grade will be horizontal, having just passed through $-4\%$ from $T_1$. Therefore, the chainage of the lowest point from the start of the curve is, by simple proportion

$$D = \frac{L}{9\%} \times 4\% = \frac{L}{A} \times g_1 \qquad\qquad (6.46)$$

which in the previous example is $\dfrac{257}{9\%} \times 4\% = 114.24$ m from $T_1$.

Knowing the chainage, the offset and the curve level at that point may be found.

This simple approach suffices as the rate of change of grade is constant for a parabola, i.e. $y = Cl^2$, $\therefore\ d^2y/dl^2 = 2C$.

### 6.11.6 Vertical-curve radius

Due to the very shallow gradients involved in VC design, the parabola may be approximated to a circular curve. In this way vertical curves may be expediently drawn on longitudinal sections using railway curves of a given radius, and vertical accelerations $(V^2/R)$ easily assessed.

In circular curves (*Section 6.1*) the main chord from $T_1$ to $T_2 = 2R \sin \Delta/2$, where $\Delta$ is the deflection angle of the two straights. In vertical curves, the main chord may be approximated to the length ($L$) of the VC and the angle $\Delta$ to the grade angle $A$, i.e.

$$\Delta \approx A\%$$
$$\therefore \quad \sin \Delta/2 \approx \Delta/2 \text{ rads} \approx A/200$$
$$\therefore \quad L \approx 2RA/200 = AR/100 \tag{6.47}$$

and as $K = L/A = R/100$, then:

$$R = 100L/A = 100K \tag{6.48}$$

It is important to note that the reduced levels of VC must always be computed. Scaling levels from a longitudinal section, usually having a vertical scale different from the horizontal, will produce a curve that is neither parabolic nor circular. The use of railway curves is simply to indicate the position and extent of the curve on the section.

Thus, having obtained the radius $R$ of the VC, it is required to know the number of the railway curve necessary to draw it on the longitudinal section.

If the horizontal scale of the section is 1 in $H$ and the vertical scale is 1 in $V$, then the number of the railway curve required to draw the VC is:

Number of railway curve in mm $= R \cdot V/H^2$, with $R$ in mm

e.g. If horizontal scale is 1 in 500, vertical scale 1 in 100 and curve radius is 200 m, then

Number of railway curve in mm $= 200\,000 \times 100/500^2 = 80$

## 6.11.7 Vertical-curve computation

The computation of a vertical curve will now be demonstrated using an example.

A '2nd difference' $(\delta^2 y/\delta l^2)$ arithmetical check on the offset computation should automatically be applied. The check works on the principle that the change of grade of a parabola $(y = C \cdot l^2)$ is constant, i.e. $\delta^2 y/\delta l^2 = 2C$. Thus, if the first and last chords are sub-chords of lengths different from the remaining standard chords, then the change of grade will be constant only for the equal-length chords.

For example, a 100-m curve is to connect a downgrade of 0.75% to an upgrade of 0.25%. If the level of the intersection point of the two grades is 150 m, calculate:

(1) Curve levels at 20-m intervals, showing the second difference $(d^2 y/dl^2)$; check on the computations.
(2) The position and level of the lowest point on the curve.

## Method

(a) Find the value of the central offset $Y$.
(b) Calculate offsets.
(c) Calculate levels along the gradients.
(d) Add/subtract (b) from (c) to get curve levels.

(a) *Referring to Figure 6.36*

Grade angle $A = (-0.75 - 0.25) = -1\%$ (this is seen automatically).

$L/2 = 50$ m, thus as the grades $IT_2$ and $IJ$ are diverging at the rate of $1\%$ (1 m per 100 m) in 50 m, then

$$T_2J = 0.5 \text{ m} = 4Y \quad \text{and} \quad Y = 0.125 \text{ m}$$

The computation can be quickly worked mentally by the student. Putting the above thinking into equation form gives

$$4Y = \frac{A}{100}\frac{L}{2} \quad \therefore \quad Y = \frac{AL}{800} \tag{6.49}$$

(b) *Offsets from equation (6.35)*

There are two methods of approach.

(1) The offsets may be calculated from one gradient throughout; i.e. $y_1$, $y_2$, $EK$, $GM$, $T_2J$, from the grade $T_1J$.
(2) Calculate the offsets from one grade, say, $T_1I$, the offsets being equal on the other side from the other grade $IT_2$.

Method (1) is preferred due to the smaller risk of error when calculating curve levels at a constant interval and grade down $T_1J$.

From equation (6.35): $\quad y_1 = Y \times \dfrac{l_1^2}{(L/2)^2}$

|  |  | 1st diff | 2nd diff |
|---|---|---|---|
| $T_1 =$ | 0 m | | |
| | | —— 0.02 | |
| $y_1 = 0.125\dfrac{20^2}{50^2} = 0.02$ m | | | —— 0.04 |
| | | —— 0.06 | |
| $y_2 = 0.125\dfrac{40^2}{50^2} = 0.08$ m | | | —— 0.04 |
| | | —— 0.10 | |
| $y_3 = 0.125\dfrac{60^2}{50^2} = 0.18$ m | | | —— 0.04 |
| | | —— 0.14 | |
| $y_4 = 0.125\dfrac{80^2}{50^2} - 0.32$ m | | | —— 0.04 |
| | | —— 0.18 | |
| $y_2 = T_2J = 4Y = 0.50$ m | | | |

The 2nd difference arithmetical check, which works only for equal chords, should be applied before any further computation.

(c) First find level at $T_1$ from known level at $I$.

Distance from $I$ to $T_1 = 50$ m, grade $= 0.75\%$ (0.75 m per 100 m)

$$\therefore \quad \text{Rise in level from } I \text{ to } T_1 = \frac{0.75}{2} = 0.375 \text{ m}$$

Level at $T_1 = 150.000 + 0.375 = 150.375$ m

Levels are now calculated at 20-m intervals along $T_1 J$, the fall being 0.15 m in 20 m. Thus, the following table may be made.

| Chainage (m) | Gradient levels | Offsets | Curve levels | Remarks |
|---|---|---|---|---|
| 0 | 150.375 | 0 | 150.375 | Start of curve $T_1$ |
| 20 | 150.225 | 0.02 | 150.245 | |
| 40 | 150.075 | 0.08 | 150.155 | |
| 60 | 149.925 | 0.18 | 150.105 | |
| 80 | 149.775 | 0.32 | 150.095 | |
| 100 | 149.625 | 0.50 | 150.125 | End of curve $T_2$ |

Position of lowest point on curve $= \dfrac{100 \text{ m}}{1\%} \times 0.75\% = 75$ m from $T_1$

$\therefore$ Offset at this point $= y_2 = 0.125 \times 75^2/50^2 = 0.281$ m

Tangent level 75 m from $T_1 = 150.375 - 0.563 = 149.812$ m

$\therefore$ Curve level $= 149.812 + 0.281 = 150.093$ m

## 6.11.8 Drawing-office practice

(1) *Design*

(a) Obtain grade angle (algebraic difference of the gradients) $A$.
(b) Extract the appropriate $K$-value from Design Table in TD 9/81.
(c) Length $(L)$ of vertical curve $= KA$.
(d) Compute offsets and levels in the usual way.

(2) *Drawing*

To select the correct railway curve for drawing the vertical curve on a longitudinal section.

(a) Find equivalent radius $(R)$ of vertical curve from $R = \dfrac{100 \cdot L}{A} = 100 \cdot K$.

(b) Number of railway curve in mm $= R$ mm $\times V/H^2$.

If horizontal scale of section is say 1/500, then $H = 500$.
If vertical scale of section is say 1/200, then $V = 200$.
If the railway curves used are still in inches, simply express $R$ in inches.

## Worked examples

*Example 6.13.* An existing length of road consists of a rising gradient of 1 in 20, followed by a vertical parabolic crest curve 100 m long, and then a falling gradient of 1 in 40. The curve joins both gradients tangentially and the reduced level of the highest point on the curve is 173.07 m above datum.

Visibility is to be improved over this stretch of road by replacing this curve with another parabolic curve 200 m long.

Find the depth of excavation required at the mid-point of the curve. Tabulate the reduced levels of points at 30-m intervals on the new curve.

What will be the minimum visibility on the new curve for a driver whose eyes are 1.05 m above the road surface? (ICE)

The first step here is to find the level of the start of the new curve; this can only be done from the information on the highest point $P$, *Figure 6.42.*

*Old curve*   $A = 7.5\%$,   $L = 100$ m

Chainage of highest point $P$ from $T_1 = \dfrac{100}{7.5\%} \times 5\% = 67$ m

Distance $T_2C$ is the divergence of the grades (7.5 m per 100 m) over half the length of curve (50) $= 7.5 \times 0.5 = 3.75$ m $= 4Y.$

$\therefore$ Central offset   $Y = 3.75/4 = 0.938$ m

Thus offset   $PB = 0.938 \cdot \dfrac{67^2}{50^2} = 1.684$ m

Therefore the level of $B$ on the tangent $= 173.07 + 1.684 = 174.754$ m. This point is 17 m from $I$, and as the new curve is 200 m in length, it will be 117 m from the start of the new curve $T_3$

$\therefore$ Fall from $B$ to $T_3$ of new curve $= 5 \times 1.17 = 5.85$ m
$\therefore$ Level of $T_3 = 174.754 - 5.85 = 168.904$ m

It can be seen that as the value of $A$ is constant, when $L$ is doubled, the value of $Y$, the central offset to the new curve, is doubled giving 1.876 m.

$\therefore$ Amount of excavation at mid-point $= 0.938$ m

| *New curve offsets* | *1st diff* | *2nd diff* |
|---|---|---|
| | —— 0.169 | |
| $y_1 = 1.876 \times \dfrac{30^2}{100^2} = 0.169$ | | —— 0.337 |
| | —— 0.506 | |
| $y_2 = 1.876 \times \dfrac{60^2}{100^2} = 0.675$ | | —— 0.339 |
| | —— 0.845 | |
| $y_2 = 1.876 \times \dfrac{90^2}{100^2} = 1.520$ | | —— 0.336 |
| | —— 1.181 | |
| $y_4 = 1.876 \times \dfrac{120^2}{100^2} = 2.701$ | | —— 0.339 |
| | —— 1.520 | |
| $y_5 = 1.876 \times \dfrac{150^2}{100^2} = 4.221$ | | —— 0.337 |
| | —— 1.857 | |
| $y_6 = 1.876 \times \dfrac{180^2}{100^2} = 6.078$ | | —— 0.431* |
| | —— 1.426 | |
| $y_7 = 4y$   $= 7.504$ | | |

\* Note change due to change in chord length from 20 m to 30 m.

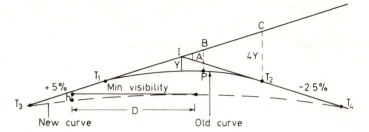

*Figure 6.42*

Levels along the tangent $T_3C$ are now obtained at 30-m intervals.

| Chainage (m) | Tangent levels | Offsets | Curve levels | Remarks |
|---|---|---|---|---|
| 0 | 168.904 | 0 | 168.904 | $T_3$ of new curve |
| 30 | 170.404 | 0.169 | 170.235 | |
| 60 | 171.904 | 0.675 | 171.229 | |
| 90 | 173.404 | 1.520 | 171.884 | |
| 120 | 174.904 | 2.701 | 172.203 | |
| 150 | 176.404 | 4.221 | 172.183 | |
| 180 | 177.904 | 6.078 | 171.826 | |
| 200 | 178.904 | 7.504 | 171.400 | $T_4$ of new curve |

From *Figure 6.42* it can be seen that the minimum visibility is half the sight distance and could thus be calculated from the necessary equation. However, if the driver's eye height of $h = 1.05$ m is taken as an offset then

$$\frac{h}{Y} = \frac{D^2}{(L/2)^2}, \quad \text{thus} \quad \frac{1.05}{1.876} = \frac{D^2}{100^2}$$

$$\therefore D = 75 \text{ m}$$

*Example 6.14.* A rising gradient $g_1$ is followed by another rising gradient $g_2$ ($g_2$ less than $g_1$). The gradients are connected by a vertical curve having a constant rate of change of gradient. Show that at any point on the curve the height $y$ above the first tangent point $A$ is given by

$$y = g_1x - \frac{(g_1 - g_2)x^2}{2L}$$

where $x$ is the horizontal distance of the point from $A$, and $L$ is the horizontal distance between the two tangent points.

Draw up a table of heights above $A$ for 100-m pegs from $A$ when $g_1 = +5\%$, $g_2 = +2\%$ and $L = 1000$ m.

At what horizontal distance from $A$ is the gradient $+3\%$? (ICE)

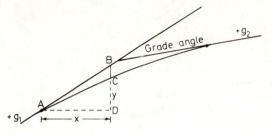

Figure 6.43

Figure 6.43 from equation for offsets.    $\dfrac{BC}{Y} = \dfrac{x^2}{(L/2)^2}$

$\therefore\ BC = Y \cdot \dfrac{4x^2}{L^2}$    but    $Y = \dfrac{AL}{8}$ stations $= \dfrac{(g_1 - g_2)L}{8}$

$\therefore\ BC = \dfrac{(g_1 - g_2)L4x^2}{8L^2} = \dfrac{(g_1 - g_2)x^2}{2L}$

Now    $BD = g_1 x$

Thus, as    $y = BD - BC = g_1 x - \dfrac{(g_1 - g_2)x^2}{2L}$

Using the above formula (which is correct only if horizontal distances $x$ and $L$ are expressed in stations, i.e. a station = 100 m)

$y_1 = 5 - \dfrac{3 \times 1^2}{20} = 4.85$ m

$y_2 = 10 - \dfrac{3 \times 2^2}{20} = 9.4$ m

$y_3 = 15 - \dfrac{3 \times 3^2}{20} = 13.65$ m    and so on

Grade angle = 3% in a 1000 m
Change of grade from 5% to 3% = 2%

$\therefore$  Distance $= \dfrac{1000}{3\%} \times 2\% = 667$ m

Example 6.15. A falling gradient of 4% meets a rising gradient of 5% at chainage 2450 m and level 216.42 m. At chainage 2350 m the underside of a bridge has a level of 235.54 m. The two grades are to be joined by a vertical parabolic curve giving 14 m clearance under the bridge. List the levels at 50-m intervals along the curve.

To find the offset to the curve at the bridge (Figure 6.44)

Level on gradient at chainage 2350 = 216.42 + 4     = 220.42 m
Level on curve at chainage    2350 = 235.54 − 14    = 221.54 m

$\therefore$  Offset at chainage        2350 = $y_2$        =    1.12 m

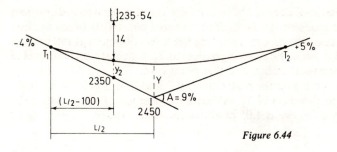

Figure 6.44

*From equation for offsets* $\dfrac{y_2}{Y} = \dfrac{(L/2 - 100)^2}{(L/2)^2}$

where $Y = \dfrac{AL}{800}$ and $A = 9\%$

$\dfrac{1.12 \times 800}{9 \times L} = \left(1 - \dfrac{200}{L}\right)^2$, and putting $x = \dfrac{200}{L}$

$1.12 \times 4x = 9(1 - x)^2$

from which $x^2 - 2.5x + 1 = 0$, giving

$x = 2$ or $0.5$ $\quad \therefore\ L = 400$ m (as $x = 2$, is not possible)

Now $Y = \dfrac{9 \times 400}{800} = 4.5$ m from which the remaining offsets are found as follows:

at chainage 50 m offset $y_1 = 4.5\ \dfrac{50^2}{200^2} = 0.28$ m

at chainage 100 m offset $y_2 = 4.5\ \dfrac{100^2}{200^2} = 1.12$ m

at chainage 150 m offset $y_3 = 4.5\ \dfrac{150^2}{200^2} = 2.52$ m

at chainage 200 m offset $Y = \phantom{xxxx} 4.50$ m

To illustrate the alternative method, these offsets may be repeated on the other gradient at 250 m $= y_3$, 300 m $= y_2$, 350 m $= y_1$. The levels are now computed along each gradient from $I$ to $T_1$ and $T_2$ respectively.

| Chainage (m) | Gradient levels | Offsets | Curve levels | Remarks |
|---|---|---|---|---|
| 0 | 224.42 | | 224.42 | Start of curve $T$ |
| 50 | 222.42 | 0.28 | 222.70 | |
| 100 | 220.42 | 1.12 | 221.54 | |
| 150 | 218.42 | 2.52 | 220.94 | |
| 200 | 216.42 | 4.50 | 220.92 | Centre of curve $I$ |
| 250 | 218.92 | 2.52 | 221.44 | |
| 300 | 221.42 | 1.12 | 222.54 | |
| 350 | 223.92 | 0.28 | 224.20 | |
| 400 | 226.42 | | 226.42 | End of curve $T_2$ |

*Example 6.16.* A vertical parabolic curve 150 m in length connects an upward gradient of 1 in 100 to a downward gradient of 1 in 50. If the tangent point $T_1$ between the first gradient and the curve is taken as datum, calculate the levels of points at intervals of 25 m along the curve until it meets the second gradient at $T_2$. Calculate also the level of the summit giving the horizontal distance of this point from $T_1$.

If an object 75 mm high is lying on the road between $T_1$ and $T_2$ at 3 m from $T_2$, and a car is approaching from the direction of $T_1$, calculate the position of the car when the driver first sees the object if his eye is 1.05 m above the road surface.          (LU)

*To find offsets*

$$A = 3\% \qquad \therefore \ 4Y = \frac{L}{200} \times 3\% = 2.25 \text{ m}$$

and                          $$Y = 0.562 \text{ m}$$

$$\therefore \ y_1 = 0.562 \times \frac{25^2}{75^2} = 0.062 \qquad y_4 = 0.562 \times \frac{100^2}{75^2} = 1.000$$

$$y_2 = 0.562 \times \frac{50^2}{75^2} = 0.250 \qquad y_5 = 0.562 \times \frac{125^2}{75^2} = 1.562$$

$$y_3 = 0.562 \times \frac{75^2}{75^2} = 0.562 \qquad y_6 = 4y = 2.250$$

Second difference checks will verify these values.

With $T_1$ at datum, levels are now calculated at 25-m intervals for 150 m along the 1 in 100 (1%) gradient.

| Chainage (m) | Gradient levels | Offsets | Curve levels | Remarks |
|---|---|---|---|---|
| 0 | 100.00 | 0 | 100.000 | Start of curve $T_1$ |
| 25 | 100.25 | 0.062 | 100.188 | |
| 50 | 100.50 | 0.250 | 100.250 | |
| 75 | 100.75 | 0.562 | 100.188 | |
| 100 | 101.00 | 1.000 | 100.000 | |
| 125 | 101.25 | 1.562 | 99.688 | |
| 150 | 101.50 | 2.250 | 99.250 | End of curve $T_2$ |

Distance of highest point from $T_1 = \dfrac{150}{3\%} \times 1\% = 50$ m

*Sight distance (S < L)*

From expression (6.37), p. 203.

$$S = [(h_1)^{1/2} + (h_2)^{1/2}]\left(\frac{200L}{A}\right)^{1/2} \quad \text{when} \quad h_1 = 1.05 \text{ m}, \quad h_2 = 0.075 \text{ m}$$

$$\therefore \ S = 130 \text{ m}, \quad \text{and the car is 17 m from } T_1 \text{ and between } T_1 \text{ and } T_2$$

*Example 6.17.* A road gradient of 1 in 60 down is followed by an up gradient of 1 in 30, the valley thus formed being smoothed by a circular curve of radius 1000 m in the

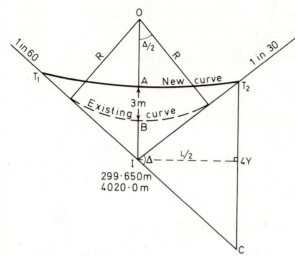

*Figure 6.45*

vertical plane. The grades, if produced, would intersect at a point having a reduced level of 299.65 m and a chainage of 4020 m.

It is proposed to improve the road by introducing a longer curve, parabolic in form, and in order to limit the amount of filling it is decided that the level of the new road at chainage 4020 m shall be 3 m above the existing surface.

Determine:

(a) The length of new curve.
(b) The levels of the tangent points.
(c) The levels of the quarter points.
(d) The chainage of the lowest point on the new curve.          (LU)

*To find central offset Y to new curve (Figure 6.45)*

From simple curve data $\Delta = \cot 60 + \cot 30 = 2° 51' 51''$

Now $BI = R(\sec \Delta/2 - 1) = 0.312$ m

∴ Central offset $AI = Y = 3.312$ m   and   $T_2C = 4Y = 13.248$ m

*To find length of new curve*

Grade 1 in 60 = 1.67%, 1 in 30 = 3.33%

∴ Grade angle $\Delta = 5\%$

(1) Then from $T_2IC$, $L/2 = \dfrac{13.248}{5} \times 100$

∴ $L = 530$ m

(2) Rise from $I$ to $T_1 = 1.67 \times 2.65 = 4.426$ m

∴ Level at $T_1 = 299.65 + 4.426 = 304.076$ m

Rise from $I$ to $T_2 = 3.33 \times 2.65 = 8.824$ m

∴ Level at $T_2 = 299.65 + 8.824 = 308.474$ m

(3) *Levels at quarter points*

1st quarter point is 132.5 m from $T_1$

∴ Level on gradient $= 304.076 - (1.67 \times 1.325) = 301.863$ m

Offset $= 3.312 \times \dfrac{1^2}{2^2} = 0.828$ m

∴ Curve level $= 301.863 + 0.828 = 302.691$ m

2nd quarter point is at 397.5 m

∴ Level on gradient $= 304.076 - (1.67 \times 3.975) = 310.714$ m

Offset $= 3.312 \times \dfrac{3^2}{2^2} = 7.452$ m

∴ Curve level $= 310.714 + 7.452 = 318.166$ m

(4) Position of lowest point on curve from $T_1 = \dfrac{530}{5\%} \times 1.67\% = 177$ m

Chainage at $T_1 = 4020 - 265 = 3755$ m

Chainage of lowest point $= 3755 + 177 = 3932$ m

## Exercises

(*6.12*) A vertical curve 120 m long of the parabola type is to join a falling gradient of 1 in 200 to a rising gradient of 1 in 300. If the level of the intersection of the two gradients is 30.36 m give the levels at 15-m intervals along the curve.

If the headlamp of a car was 0.375 m above the road surface, at what distance will the beam strike the road surface when the car is at the start of the curve ? Assume the beam is horizontal when the car is on a level surface. (LU)

(*Answer:* 30.660, 30.594, 30.504, 30.486, 30.477, 30.489, 30.516, 30.588; 103.8 m)

(*6.13*) A road having an up gradient of 1 in 15 is connected to a down gradient of 1 in 20 by a vertical parabolic curve 120 m in length. Determine the visibility distance afforded by this curve for two approaching drivers whose eyes are 1.05 m above the road surface.

As part of a road improvement scheme a new vertical parabolic curve is to be set out to replace the original one so that the visibility distance is increased to 210 m for the same height of driver's eye.

Determine:

(a) The length of new curve.
(b) The horizontal distance between the old and new tangent points on the 1 in 5 gradient.
(c) The horizontal distance between the summits of the two curves.

(ICE)

(*Answer:* 92.94 m, (a) 612 m, (b) 246 m, (c) 35.7 m)

(*6.14*) A vertical parabolic sag curve is to be designed to connect a down gradient of 1 in 20 with an up gradient of 1 in 15, the chainage and reduced level of the intersection point of the two gradients being 797.7 m and 83.544 m respectively.

In order to allow for necessary headroom, the reduced level of the curve at chainage 788.7 m on the down gradient side of the intersection point is to be 85.044 m.

Calculate:

(a) The reduced levels and chainages of the tangent points and the lowest point on the curve.

(b) The reduced levels of the first two pegs on the curve, the pegs being set at the 30-m points of through chainage. (ICE)

(*Answer:* $T_1 = 745.24$ m, 86.166 m, $T_2 = 850.16$ m, 87.042 m, lowest pt = 790.21 m, 85.041 m, (b) 85.941 m, 85.104 m)

(*6.15*) The surface of a length of a proposed road consists of a rising gradient of 2% followed by a falling gradient of 4% with the two gradients joined by a vertical parabolic summit curve 120 m in length. The two gradients produced meet a reduced level of 28.5 m OD.

Compute the reduced levels of the curve at the ends, at 30-m intervals and at the highest point.

What is the minimum distance at which a driver, whose eyes are 1.125 m above the road surface, would be unable to see an obstruction 100 mm high? (ICE)

(*Answer:* 27.300, 27.675, 27.600, 27.075, 26.100 m; highest pt, 27.699 m, 87 m)

# Underground and hydrographic surveying

## 7.1 UNDERGROUND SURVEYING

The essential problem in underground surveying is that of orientating the underground surveys to the surface surveys, the procedure involved being termed a *correlation*.

In an underground transport system, for instance, the tunnels are driven to connect inclined or vertical shafts (points of surface entry to the transport system) whose relative locations are established by surface surveys. Thus the underground control networks must be connected and oriented into the same co-ordinate system as the surface networks. To do this, one must obtain the co-ordinates of at least one underground control station and the bearing of at least one line of the underground network, relative to the surface network.

Another prime example of orientation is that of the National Coal Board (NCB) of the UK, all of whose underground surveys are required, by law, to be orientated and connected to the Ordnance Survey national grid (NG) system.

If entry to the underground tunnel system is via an inclined shaft, then the surface survey may simply be extended and continued down that shaft and into the tunnel, usually by the method of traversing. Extra care would be required in the measurement of the horizontal angles due to the steeply-inclined sights involved (refer *Section 4.2.2*) and in the temperature corrections to the taped distances due to the thermal gradients encountered.

If entry is via a vertical shaft, then optical, mechanical or gyroscopic methods of orientation are used.

### 7.1.1 Optical methods

Where the shaft is shallow and of relatively large diameter the bearing of a surface line may be transferred to the shaft bottom by theodolite (*Figure 7.1*).

The surface stations *A* and *B* are part of the control system and represent the direction in which the tunnel must proceed. They would usually be established clear of ground movement caused by shaft sinking or other factors. Auxiliary stations *c* and *d* are very carefully aligned with *A* and *B* using the theodolite on both faces and with due regard to all error sources. The relative bearing of *A* and *B* is then transferred to *A'B'* at

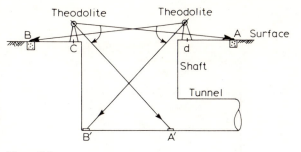

*Figure 7.1*

the shaft bottom by direct observations. Once again these observations must be carried out on both faces with the extra precautions advocated for steep sights.

If the co-ordinates of $d$ are known then the co-ordinates of $B'$ could be fixed by measuring the vertical angle and distance (EDM) to a reflector at $B'$.

It is important to understand that the accurate orientation bearing of the tunnel is infinitely more critical than the co-ordinate position. For instance, a standard error of $\pm 1'$ in transferring the bearing down the shaft to $A'B'$ would result in a positional error at the end of 1 km of tunnel drivage of $\pm 300$ mm and would increase to $\pm 600$ mm after 2 km of drivage.

## 7.1.2 Mechanical methods

Although these methods, which involve the use of wires hanging vertically in a shaft, are rapidly being superseded by gyroscopic methods, they are still widely used and are described herewith.

The basic concept is that wires hanging freely in a shaft will occupy the same position underground that they do at the surface, hence the bearing of the wire plane will remain constant throughout the shaft.

### 7.1.2.1 Weisbach triangle method

This appears to be the most popular method in civil engineering. Two wires, $W_1$ and $W_2$, are suspended vertically in a shaft forming a very small base line (*Figure 7.2*). The principle is to obtain the bearing and co-ordinates of the wire base relative to the surface base. These values can then be transferred to the underground base.

In order to establish the bearing of the wire base at the surface, it is necessary to compute the angle $W_s W_2 W_1$ in the triangle as follows:

$$\sin \hat{W}_2 = \frac{w_2}{w_s} \sin \hat{W}_s \tag{7.1}$$

As the Weisbach triangle is formed by approximately aligning the Weisbach station $W_s$ with the wires, the angles at $W_s$ and $W_2$ are very small and equation (7.1) may be written:

$$\hat{W}_2'' = \frac{w_2}{w_s} \hat{W}_s'' \tag{7.2}$$

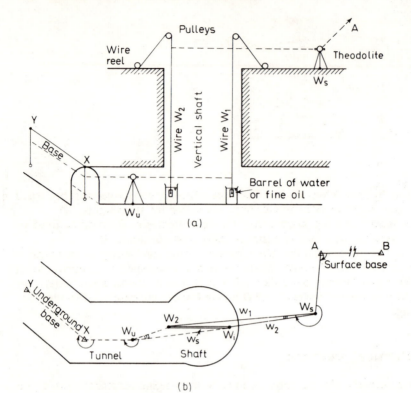

Figure 7.2 (a) Section, and (b) plan

(The expression is accurate to seven figures when $\hat{W}_s < 18'$ and to six figures when $\hat{W}_s < 45'$.)

From equation (7.2), it can be seen that the observational error in angle $W_s$ will be multiplied by the fraction $w_2/w_s$.

Its effect will therefore be reduced if $w_2/w_s$ is less than unity. Thus the theodolite at $W_s$ should be as near the front wire $W_1$ as focusing will permit and preferably at a distance smaller than the wire base $W_1 W_2$.

The following example, using simplified data, will now be worked to illustrate the procedure. With reference to Figure 7.2(b), the following field data is obtained:

(1) Surface observations

| | |
|---|---|
| Angle $BAW_s$ = 90° 00′ 00″ | Distance $W_1 W_2 = w_s = 10.000$ m |
| Angle $AW_s W_2$ = 260° 00′ 00″ | Distance $W_1 W_s = w_2 = 5.000$ m |
| Angle $W_1 W_s W_2$ = 0° 01′ 20″ | Distance $W_2 W_s = w_1 = 15.000$ m |

(2) Underground observations

| | |
|---|---|
| Angle $W_2 W_s W_1$ = 0° 01′ 50″ | Distance $W_2 W_u = y = 4.000$ m |
| Angle $W_1 W_u X$ = 200° 00′ 00″ | Distance $W_u W_1 = x = 14.000$ m |
| $W_u XY$ = 240° 00′ 00″ | |

Solution of the surface Weisbach triangle:

$$\text{Angle } W_s W_2 W_1 = \frac{5}{10} \times 80'' = 40''$$

Similarly, underground

$$\text{Angle } W_2 W_1 W_u = \frac{4}{10} \times 110'' = 44''$$

The bearing of the underground base $XY$, relative to the surface base $AB$ is now computed in a manner similar to a traverse:

| | | |
|---|---|---|
| Assuming | WCB of $AB$ | $= \quad 89° \ 00' \ 00''$ |
| then, | WCB of $AW_s$ | $= 179° \ 00' \ 00''$   (using angle $BAW_s$) |
| | Angle $AW_s W_2$ | $= 260° \ 00' \ 00''$ |
| | | sum $= 439° \ 00' \ 00''$ |
| | | $-180°$ |
| | WCB of $W_s W_2$ | $= 259° \ 00' \ 00''$ |
| the reverse bearing $W_2 W_s$ | | $= \quad 79° \ 00' \ 00''$ |
| | Angle $W_s W_2 W_1$ | $= +0° \ 00' \ 40''$ |
| | WCB $W_2 W_1$ | $= \quad 79° \ 00' \ 40''$   (wire base) |
| reverse | WCB $W_1 W_2$ | $= 259° \ 00' \ 40''$ |
| | Angle $W_2 W_1 W_u$ | $= -0° \ 00' \ 44''$ |
| | WCB $W_1 W_u$ | $= 258° \ 59' \ 56''$ |
| | Angle $W_1 W_u X$ | $= 200° \ 00' \ 00''$ |
| | | sum $= 458° \ 59' \ 56''$ |
| | | $-180°$ |
| | WCB $W_u X$ | $= 278° \ 59' \ 56''$ |
| | Angle $W_u XY$ | $= 240° \ 00' \ 00''$ |
| | | sum $= 518° \ 59' \ 56''$ |
| | | $-180°$ |
| | WCB $XY$ | $= 338° \ 59' \ 56''$   (underground base) |

The transfer of bearing is of prime importance; the co-ordinates can be obtained in the usual way by incorporating all the measured lengths $AB$, $AW_s$, $W_u X$, $XY$.

### 7.1.2.2 Shape of the Weisbach triangle

As already indicated, the angles $W_2$ and $W_s$ in the triangle are as small as possible. The reason for this can be illustrated by considering the effect of accidental observation errors on the computed angle $W_2$.

From the basic equation: $\sin \hat{W}_2 = \dfrac{w_2}{w_s} \sin \hat{W}_s$

differentiate with respect to each of the measured quantities in turn:

(1) with respect to $W_s$

$$\cos W_2 \, \delta W_2 = \frac{w_2}{w_s} \cos W_s \, \delta W_s$$

$$\therefore \ \delta W_2 = \frac{w_2 \cos W_s}{w_s \cos W_2} \delta W_s$$

(2) with respect to $w_2$

$$\cos W_2 \, \delta W_2 = \frac{\sin W_s}{w_s} \delta w_2$$

$$\therefore \ \delta W_2 = \frac{\sin W_s}{w_s \cos W_2} \delta w_2$$

(3) with respect to $w_s$

$$\cos W_2 \, \delta W_2 = \frac{-w_2 \sin W_s}{w_s^2} \delta w_s$$

$$\therefore \ \delta W_2 = \frac{-w_2 \sin W_s}{w_s^2 \cos W_2} \delta w_s$$

then:

$$\delta W_2 = \pm \left[ \frac{w_2^2 \cos^2 W_s}{w_s^2 \cos^2 W_2} \delta W_s^2 + \frac{\sin^2 W_s}{w_s^2 \cos^2 W_2} \delta w_2^2 + \frac{w_2^2 \sin^2 W_s}{w_s^4 \cos^2 W_2} \delta w_s^2 \right]^{1/2}$$

$$= \pm \frac{w_2}{w_s \cos W_2} \left[ \cos^2 W_s \, \delta W_s^2 + \sin^2 W_s \frac{\delta w_2^2}{w_2^2} + \sin^2 W_s \frac{\delta w_s^2}{w_s^2} \right]^{1/2}$$

but $\cos W_s = \dfrac{\sin W_s \cos W_s}{\sin W_s} = \sin W_s \cot W_s$, which on substitution gives:

$$\delta W_2 = \pm \frac{w_2}{w_s \cos W_2} \left[ \sin^2 W_s \cot^2 W_s \, \delta W_s^2 + \sin^2 W_s \frac{\delta w_2^2}{w_2^2} + \sin^2 W_s \frac{\delta w_s^2}{w_s^2} \right]^{1/2}$$

$$= \pm \frac{w_2 \sin W_s}{w_s \cos W_2} \left[ \cot^2 W_s \, \delta W_s^2 + \frac{\delta w_2^2}{w_2^2} + \frac{\delta w_s^2}{w_s^2} \right]^{1/2}$$

by sine rule $\dfrac{w_2 \sin W_s}{w_s} = \sin W_2$, therefore substituting

$$\delta W_2 = \pm \tan W_2 \left[ \cot^2 W_s \, \delta W_s^2 + \left( \frac{\delta w_2}{w_2} \right)^2 + \left( \frac{\delta w_s}{w_s} \right)^2 \right]^{1/2} \tag{7.3}$$

Thus to reduce the standard error $(\delta W_2)$ to a minimum:

(1) $\tan W_2$ must be a minimum; therefore the angle $W_2$ should approach $0°$.
(2) As $W_2$ is very small, $W_s$ will be very small and so $\cot W_s$ will be very large. Its effect will be greatly reduced if $\delta W_s$ is very small; the angle $W_s$ must therefore be measured with maximum precision.

### 7.1.2.3 Sources of error

The standard error of the transferred bearing $e_B$, is the combined effect of:

(1) Errors in connecting the surface base to the wire base, $e_s$.
(2) Errors in connecting the wire base to the underground base, $e_u$.
(3) Errors in the determination of the verticality of the wire plane, $e_p$ giving:

$$e_B = \pm (e_s^2 + e_u^2 + e_p^2)^{1/2} \qquad (7.4)$$

The errors $e_s$ and $e_u$ can be obtained in the usual way from an examination of the method and type of instruments used. The source of error $e_p$ is vitally important in view of the extremely short length of the wire base.

Given random errors $e_1$ and $e_2$, of deflected wires $W_1$ and $W_2$ equal to 1 mm, then $e_p = 100$ sec for a wire base of 2 m. The NCB specifies a value of $2'\,00''$ for $e_B$, then from equation (7.4) assuming $e_s = e_u = e_p$:

$$e_p = \frac{2'\,00''}{(3)^{1/2}} = 70'',$$

which for the same wire base of 2 m permits a deflection of the wires of only 0.7 mm. These figures serve to indicate the great precision and care needed in plumbing a shaft, in order to minimize orientation errors.

### 7.1.2.4 Verticality of the wire plane

The factors affecting the verticality of the wires are:

(1) *Ventilation air currents in the shaft*
   All forced ventilation should be shut off and the plumb bob protected from natural ventilation.
(2) *Pendulous motion of the shaft plumb*
   The motion of the plumb bob about its suspension point can be reduced by immersing it in a barrel of water or fine oil. When the shaft is deep, complete elimination of motion is impossible and clamping of the wires in their mean swing position may be necessary.
   The amplitude of wire vibrations, which induce additional motion to the swing, may be reduced by using a heavy plumb bob, with its point of suspension close to the centre of its mass, and fitted with large fins.
(3) *Spiral deformation of the wire*
   Storage of the plumb wire on small diameter reels gives a spiral deformation to the wire. Its effect is reduced by using a plumb bob of maximum weight. This should be calculated for the particular wire using a reasonable safety factor.

These sources of error are applicable to all wire surveys.

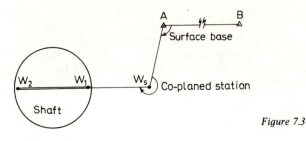

Figure 7.3

### 7.1.2.5 Co-planing

The principles of this alternative method are shown in *Figure 7.3*. The triangle of the previous method is eliminated by aligning the theodolite at $W_s$ exactly with the wires $W_1$ and $W_2$. This alignment is easily achieved by trial and error, focusing first on the front wire and then on the back. Both wires can still be seen through the telescope even when in line. The instrument should be set up within 3 to 4 m of the nearer wire. Special equipment is available to prevent lateral movement of the theodolite affecting its level, but if this is not used, special care should be taken to ensure that the tripod head is level.

The movement of the focusing lens in this method is quite long. Thus for alignment to be exact, the optical axis of the object lens should coincide exactly with that of the focusing lens in all focusing positions. If any large deviation exists, the instrument should be returned to the manufacturer. The chief feature of this method is its simplicity with little chance of gross errors.

### 7.1.2.6 Weiss quadrilateral

This method may be adopted when it is impossible to set up the theodolite, even approximately, on the line of the wire base $W_1 W_2$ (*Figure 7.4*). Theodolites are set up at $C$ and $D$ forming a quadrilateral $CDW_1W_2$. The bearing and co-ordinates of $CD$ are obtained relative to the surface base, the orientation of the wire base being obtained through the quadrilateral. Angles 1, 2, 3 and 8 are measured direct, angles 4 and 7 are

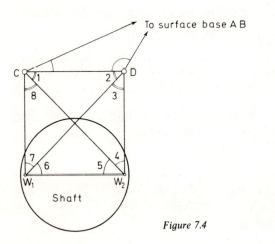

Figure 7.4

obtained as follows:

Angle $4 = [180° - (\hat{1} + \hat{2} + \hat{3})]$

Angle $7 = [180° - (\hat{1} + \hat{2} + \hat{8})]$

The remaining angles 6 and 5 are then computed from

$$\sin\hat{1}\sin\hat{3}\sin\hat{5}\sin\hat{7} = \sin\hat{2}\sin\hat{4}\sin\hat{6}\sin\hat{8}$$

thus
$$\frac{\sin\hat{5}}{\sin\hat{6}} = \frac{\sin\hat{2}\sin\hat{4}\sin\hat{8}}{\sin\hat{1}\sin\hat{3}\sin\hat{7}} = x \qquad\qquad (a)$$

and
$$(\hat{5} + \hat{6}) = (\hat{1} + \hat{2}) = \hat{y}$$

$$\therefore\ \hat{5} = (\hat{y} - \hat{6}) \qquad\qquad (b)$$

from (a) $\sin\hat{5} = x\sin\hat{6}$    $\therefore\ \sin(\hat{y} - \hat{6}) = x\sin\hat{6}$

and    $\sin\hat{y}\cos\hat{6} - \cos\hat{y}\sin\hat{6} = x\sin\hat{6}$

from which $\sin y\cot\hat{6} - \cos\hat{y} = x$

and
$$\cot\hat{6} = \frac{x + \cos\hat{y}}{\sin\hat{y}} \qquad\qquad (7.5)$$

Having found angle 6 from equation (7.5), angle 5 is found by substitution in (b). Error analysis of the observed figure indicates

(1) The best shape for the quadrilateral is a square.
(2) Increasing the ratio of the length $CD$ to the wire base increases the standard error of orientation.

## 7.1.2.7 Single wires in two shafts

The above methods have dealt with orientation through a single shaft, which is the general case in civil engineering. Where two shafts are available, orientation can be achieved via a single wire in each shaft. This method gives a longer wire base, and wire deflection errors are much less critical.

The principles of the method are outlined in *Figure 7.5*. Single wires are suspended in each shaft at $A$ and $B$ and co-ordinated into the surface control network, most probably by multiple intersection from as many surface stations as possible. From the co-ordinates of $A$ and $B$, the bearing $AB$ is obtained.

A traverse is now carried out from $A$ to $B$ via an underground connecting tunnel (*Figure 7.5(a)*). However, as the angles at $A$ and $B$ cannot be measured it becomes an open traverse on an *assumed* bearing for $AX$. Thus, if the assumed bearing for $AX$ differed from the 'true' (but unknown) bearing by $\alpha$, then the whole traverse would swing to apparent positions $X'$, $Y'$, $Z'$ and $B'$ (*Figure 7.5(b)*).

The value of $\alpha$ is the difference of the bearings $AB$ and $AB'$ computed from surface and underground co-ordinates respectively. Thus if the underground bearings are rotated by the amount $\alpha$ this will swing the traverse almost back to $B$. There will still be a small misclosure due to linear error and this can be corrected by multiplying each length by a scale factor equal to length $AB$/length $AB'$. Now, using the corrected bearings and lengths the corrected co-ordinates of the traverse fitted to $AB$ can be calculated. These co-ordinates will be relative to the surface co-ordinate system.

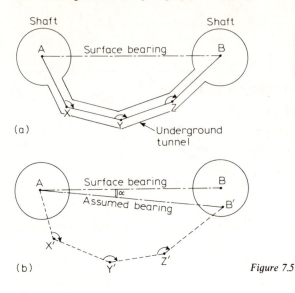

*Figure 7.5*

Alternatively, the corrected co-ordinates can be obtained directly by mathematical rotation and translation of the 'assumed' values. Therefore, with $A$ as origin, a rotation equal to $\alpha$ and a translation of $AB/AB'$, the corrected co-ordinates are obtained from

$$E_i = E_0 + K(E_i' \cos \alpha - N_i' \sin \alpha) \qquad (7.6a)$$

$$N_i = N_0 + K(N_i' \cos \alpha + E_i' \sin \alpha) \qquad (7.6b)$$

where   $E_0, N_0$ = co-ordinates of the origin (in this case $A$)
  $E_i', N_i'$ = co-ordinates of the traverse points computed on the assumed bearing
  $E_i, N_i$ = transformed co-ordinates of the underground traverse points
  $K$ = scale factor (length $AB$/length $AB'$)

There is no doubt that this is the most accurate and reliable method of surface-to-underground orientation. The accuracy of the method is dependent upon

(a) The accuracy of fixing the position of the wires at the surface.
(b) The accuracy of the underground connecting traverse.

The influence of errors in the verticality of the wires, so critical in single-shaft work, is practically negligible owing to the long distance separating the two shafts. Provided that the legs of the underground traverse are long enough, then modern single-second theodolites integrated with EDM equipment would achieve highly accurate surface and underground surveys, resulting in final orientation accuracies of a few seconds. As the whole procedure is under strict control there is no reason why the final accuracy cannot be closely predicted.

### 7.1.2.8 Alternatives

In all the above methods the wires could be replaced by autoplumbs or lasers.
  In the case of the autoplumb, stations at the shaft bottom could be projected

vertically up to specially arranged targets at the surface and appropriate observations taken direct on to these points.

Similarly, lasers could be arranged at the surface to project the beam vertically down the shaft to be picked up on optical or electronic targets. The laser spots then become the shaft stations correlated in the normal way.

The major problems encountered by using the above alternatives are:

(1) Ensuring the laser beam is vertical.
(2) Ensuring correct detection of the centre of the beam, and
(3) Refraction in the shaft (applies also to autoplumb).

In the first instance a highly-sensitive spirit level or automatic compensator could be used; excellent results have been achieved using arrangements involving mercury pools, or photo-electric sensors.

Detecting the centre of the laser is more difficult. Lasers having a divergence of 10″ to 20″ would give a spot of 10 mm and 20 mm diameter respectively at 200 m. This spot also tends to move about due to variations in air density. It may therefore require an arrangement of photocell detectors to solve the problem.

Due to the turbulence of air currents in shafts the problem of refraction has not proved too dangerous.

## 7.2 GYRO-THEODOLITE

An alternative to the use of wire methods is the gyro-theodolite. This is a north-seeking gyroscope integrated with a theodolite, and can be used to orient underground base lines relative to true north.

The main type is the suspended gyroscope used by the Wild GAK.1. The essential elements of the suspended gyro-theodolite are shown in *Figure 7.6*.

### 7.2.1 Theory

The gyroscope is basically a rapidly-spinning flywheel with the spin axis horizontal. The gyro spins from west to east, as does the Earth, the horizontal component of the Earth's rotation causing the spin axis to oscillate about the true north position.

Before commencing an explanation of the theory of the north-seeking gyroscope, a revision of Newton's Laws of Motion may prove useful. If the force $F$ increases the velocity of a mass $m$ from $V_1$ to $V_2$ in time $t$ then

$$F \propto (mV_2 - mV_1)/t$$

as $(V_2 - V_1)/t = $ acceleration $= a$, then $F \propto ma$.

The constant of proportionality $C$ in the equation $F = Cma$ can be made unity by suitably defining the units of $F$. Using SI units, $C$ does in fact become unity.

$$\therefore \quad F = (mV_2 - mV_1)/t = \text{rate of change of linear momentum.}$$

Similarly,   $T = (I\Omega_2 - I\Omega_1)/t = $ rate of change of angular momentum

thus   $T \propto F$   ($I = $ moment of inertia, $\Omega = $ angular velocity of spin)

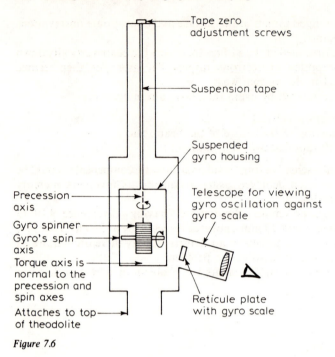

Tape zero
adjustment screws

Suspension tape

Suspended
gyro housing

Precession
axis

Telescope for viewing
gyro oscillation against
gyro scale

Gyro spinner

Gyro's spin
axis

Torque axis is
normal to the
precession and
spin axes

Reticule plate
with gyro scale

Attaches to top
of theodolite

*Figure 7.6*

The theory may be itemized as follows:

(1) *Figure 7.7(a)* indicates a spinning flywheel in which the angular velocity of spin $= \Omega$.
(2) The angular velocity of spin results in an angular momentum vector (AMV) $OA$ (similar to that on a RH screw as it enters).
(3) Consider now the AMV changing position to $OB$ in the horizontal plane $AOB$ during time $t$.
(4) This results in a change in angular momentum of $AB = (I\Omega_2 - I\Omega_1)$, which for a small displacement may be regarded as a vector change at 90° to $OA$.
(5) For the angular momentum vector to change position from $A$ to $B$, an additional vector quantity must be superimposed on the system. Such a quantity is the reactive torque effect $T$ along the axis parallel to $AB$ (torque axis). Thus $T = (I\Omega_2 - I\Omega_1)/t$.
(6) A force $F$ acting vertically down on the spin axis will produce the reactive torque effect $T$, i.e. $T = FR$.

Thus to summarize:

The effect of a force $F$ acting vertically down on the spin axis of a spinning gyro, is to cause the spin axis to *precess* in a horizontal plane about the vertical axis of precession.

Precession will continue, and resistance to the couple likewise, until the plane of the gyro rotor coincides with the plane of the applied couple. Precession then ceases and with it all resistance to the applied couple.

The effect of the Earth's rotation on the gyro fulfills the stated summary as follows:

Consider *Figure 7.7(b)* with the gyro at 0h having its spin axis E–W. Due to the phenomenon of gyroscopic inertia it will maintain its plane of rotation in space while

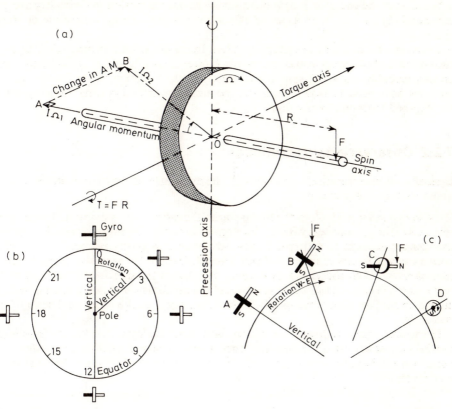

Figure 7.7

the Earth's horizon turns in space. Thus, although maintaining its plane, it appears to turn with respect to the Earth at the rate of one revolution per 24 hours.

At *A* (*Figure 7.7(c)*) consider a weight in the form of a pendulum attached to the gyro axle, it will point to the centre of the Earth, and render the spin axis horizontal. Assume the axis is E–W.

At *B*, the Earth's rotation causes the axis to show an apparent tilt as described in *Figure 7.7(b)*. The pendulum weight is now no longer evenly supported so that the effect of gravity is felt mainly by the upper end of the axis. The effect of this downward force *F* is to cause precession as shown at *C*.

At *D*, precession has swung the spin axis into the N–S meridian. In this position the direction of the rotor spin is the same as the Earth's and so has no effect on the pendulum weight. Thus, theoretically, the spin axis will point N–S and all movement ceases. In practice, however, the inertia of the system causes the spin axis to over-shoot the N–S meridian which results in oscillations of the spin axis about the meridian.

From the theory of the spinning wheel, provided that the total angular rotation is small, the motion of the horizontal spin axis about the vertical may be represented by:

$$K_1\ddot{\theta} + K_2\dot{\theta} + K_3\theta = 0 \qquad (7.7)$$

where $\theta$ = angle between spin axis and true north, and $K_1, K_2, K_3$ are constants.

Solution of the equation is a damped simple harmonic motion in which $\theta$ converges exponentially to zero, with a period of several minutes and an amplitude of about $\pm 2.5°$.

Location of true north, therefore, involves fixing the axis of symmetry of the gyro oscillation as it precesses about true north. However, as it would be time-wasting and uneconomical to allow the spin axis to steady in the direction of true north, various methods have been devised to compute the necessary direction from observations of the damped harmonic oscillation.

### 7.2.2 Observational techniques

Basically, all the techniques used measure the amplitude or the time of the oscillation about its axis of symmetry.

(1) *Reversal-point method:* In the first instance the instrument is orientated so that the spin axis of the gyroscope is within a few degrees of north. The power is switched on and the spinner brought to full speed, at which point it is gently uncaged; it will then oscillate to and fro about its vertical axis. The oscillation is then tracked, using the slow-motion screws of the theodolite, by keeping the image of the spin axis in the centre of the 'gyro scale' (*Figure 7.8(b)*). The magnitude of the oscillation, which is a damped simple harmonic motion about true north, is then measured on the horizontal scale of the theodolite.

Horizontal circle readings are taken each time the gyro reaches its maximum oscillation east or west of the meridian; these positions ($r$) are called *reversal points*. The minimum number of readings required is three, the mean of which gives the direction of true north ($N$).

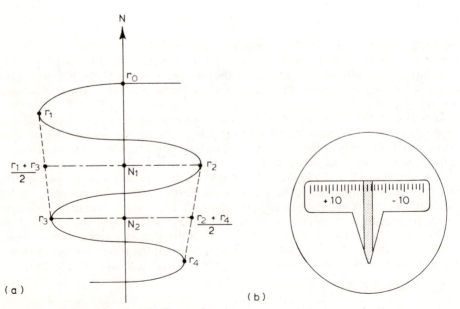

Figure 7.8 (a) Gyro precession. (b) Gyro-scale

Various methods exist for finding the mean, the most popular being Schuler's Mean, which is explained by reference to *Figure 7.8(a)*.

$$N_1 = \frac{1}{2}\left(\frac{r_1 + r_3}{2} + r_2\right) = \frac{1}{4}(r_1 + 2r_2 + r_3) \tag{7.8}$$

an additional observation $r_4$, enables a second mean to be computed, thus:

$$N_2 = \frac{1}{2}\left(\frac{r_2 + r_4}{2} + r_3\right) = \frac{1}{4}(r_2 + 2r_3 + r_4) \tag{7.9}$$

The direction of true north $(N)$ is then more accurately obtained from $N = \frac{1}{2}(N_1 + N_2)$.

Dr T. L. Thomas (*Proc., 3rd South African Nat. Survey Conf.*, January 1967) advocates the symmetrical four-point method, using two observations to the east, and two to the west. This formula is simply the mean of $(N_1 + N_2)$

i.e.   $N = (r_1 + 3r_2 + 3r_3 + r_4)/8$ \tag{7.10}

In this method, as the theodolite moves with the spinner, the suspension tape should not twist and untwist with reference to it, hence there should be no tape zero error. However, tests have proved that a small tape zero error due to torque in the suspension system, does require a correction (refer *Section 7.2.3(2)*). Thus, if

$N$ = the horizontal circle reading of gyro north
$Z$ = the tape zero correction, and
$N_0$ = the horizontal circle reading of true north,

then   $N_0 = (N - Z)$ \tag{7.11}

Further corrections also are required for the instrument constant (refer *Section 7.2.3(1)*).

(2) *Transit method:* This method, as originally devised for the Wild GAK.1 gyro-theodolite, assumed the damping effect was zero and that within $\pm 15'$ of north $(N')$ the oscillation curve was linear and hence proportional to time.

In this method the theodolite remains clamped and the magnitude of the oscillation is measured against the gyro scale, as shown in *Figure 7.9*. As the moving mark, which depicts the oscillation, passes through the central graduation of the gyro scale, its time is noted by means of a stop watch (with lap-timing facilities).

From *Figure 7.9*, as the moving mark passes the zero of the gyro scale, the time is zero $(t_1)$ and the stop watch is started. When it reaches its western elongation its scale reading is noted $(A_W)$. As it returns to the zero mark, the time $(t_2)$ is noted. Thus $t_2 - t_1 = T_W$, is the time taken for the gyro to reach its western reversal point and return and is called the *half oscillation time of the western elongation*. The eastern elongation reading $(A_E)$ is noted and the return transit time $t_3$, from which $t_3 - t_2 = T_E$.

The correction $\Delta N$, to transform the approximate north reading $N'$ to true north $N$, is given by

$$\Delta N = CA\Delta T \tag{7.12}$$

where    $A$ = the amplitude = $(A_W + A_E)/2$
$C$ = the proportionality constant, which changes with change in latitude
$\Delta T$ = the difference in 'swing' time = algebraic sum of times $T_E$ and $T_W$ (by convention $T_E$ is positive and $T_W$ negative)

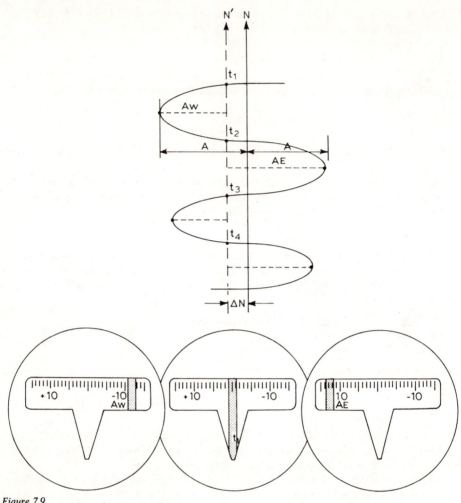

*Figure 7.9*

It is obvious from the diagrams that if $A_E > A_W$, then $N'$ is west of $N$ and the correction $\Delta N$ is positive, and *vice versa*.

The value of $C$ need be obtained only once for each instrument, as follows. With alidade oriented first east and then west of true north, the usual transit observations are taken, giving two equations:

$$N = N'_1 + CA_1\Delta T_1$$
$$N = N'_2 + CA_2\Delta T_2$$

then   $C = \dfrac{N'_1 - N'_2}{A_2\Delta T_2 - A_1\Delta T_1}$   mins of arc/div of scale/sec of time.                    (7.13)

As this method has the theodolite in the clamped mode, the harmonic motion of the spin axis is affected by torque due to precession and torque due to twisting and untwisting of the suspension system.

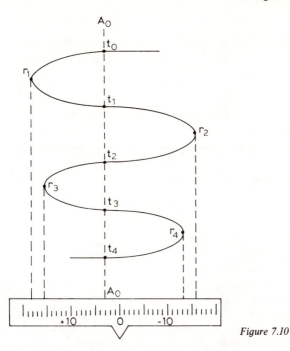

Figure 7.10

Also, in the case of the Wild GAK.1, the gyro scale intervals are 10′ and therefore cannot be read to the accuracy inherent in the gyro itself. The Royal School of Mines (RSM) recommended the introduction of an optical micrometer which enabled readings to 1/100 of the scale readings with an accuracy of about ±4″.

(3) *Transit method using the Wild modified GAK.1:* This method (*Figure 7.10*) was devised at the RSM and involves timing the moving mark as it passes *any* graduation on the gyro scale, i.e. time $t_0$ at graduation $A_0$, then reading the reversal point $r_1$, and then recording the time $t_1$ when the mark returns to the selected gyro scale graduation $A_0$. This process is continued until sufficient readings are available. The value $N$, of the centre of the oscillations, may be found as follows:

$$T = \text{period of oscillation} = \frac{1}{3}(t_2 - t_0) + (t_3 - t_1) + (t_4 - t_2) \tag{7.14}$$

$$u = \frac{90}{T}(t_1 - t_0) \tag{7.15}$$

$$P = 2\sin^2 u \tag{7.16}$$

$$N = \frac{A_0 + r_1(P - 1)}{P} \tag{7.17}$$

Further values of $T$ can be obtained for different values of $A_0$.

(4) *The amplitude method:* This method (*Figure 7.10*) was also devised at the RSM and uses the modified instrument in the clamped mode within a few degrees of true north. The improved accuracy of reading on the gyro scale enables reversal point readings $(r_n)$ to be made satisfactorily, then

$$N = (r_1 + 3r_2 + 3r_3 + r_4)/8$$

and if $F$ = the fixed reading on the horizontal circle of the theodolite at the period of observation then the horizontal circle reading of true north $N_0$ is obtained from

$$N_0 = F - sN(1 + C) - Z \qquad (7.18)$$

where   $C$ = proportionality constant
   $Z$ = tape zero correction $(Csr_0)$ (refer *Section 7.2.3(2)*)
   $s$ = the value of 1 div on the gyro scale.

It can be seen that methods (3) and (4) could be (and usually are) combined. Other methods of observation exist and may be referred to in *The Six Methods of Finding North Using a Suspended Gyroscope* by Dr T. L. Thomas (*Survey Review*, Vol. 26, 203, January 1982 and 204, April 1982).

### 7.2.3 Instrumental errors

### (1)  Instrument calibration constant

The scale on which the oscillation of the gyroscope is observed, as in the amplitude method, is usually not exactly aligned with the north seeking vector of the gyroscope. Also, the line defined by the gyro scale may not be aligned with the axis of the theodolite. These errors therefore constitute an instrument constant $K$ which can be ascertained only by carrying out observations on a base line of known azimuth.

Then   $N = N_G + K$

where   $N$ = true or geographical north
   $N_G$ = gyro north (i.e. the apparent $N$ established by the gyroscope)
   $K$ = instrument calibration constant

Thus:   Azimuth of known base =   30° 25′ 30″
   Gyro azimuth of base   =   30° 28′ 30″
   _____
   $K$-value        = −0° 03′ 00″

Tests have shown that the $K$-value is not a constant but changes slowly over a period of time. Frequent calibration checks are therefore necessary to obtain optimum results when using a gyro-theodolite, and should certainly be carried out just before and immediately after underground observations.

### (2)  Tape zero error

The tape zero position is defined as *the position of rest of the oscillating system relative to the instrument*, with the gyro in the non-spin mode. In the zero position any torquing forces on the suspension tape holding the gyro-system, due to twist in the tape, are eliminated and the gyro mark coincides with the zero of the gyro scale.

In order to find the tape zero correction $(Z)$, the gyro oscillations $(r_n)$ in the non-spin mode are read against the gyro scale as follows:

$$r_B = (r_1 + 3r_2 + 3r_3 + r_4)/8 \quad \text{(amplitude readings before spin up)}$$
$$r_A = (r'_1 + 3r'_2 + 3r'_3 + r'_3)/8 \quad \text{(amplitude readings after spin up)}$$

then the weighted mean $r_0$ is formed from

$$r_0 = (r_B + 3r_A)/4 \tag{7.19}$$

The correction $(Z)$ to normal north-finding observations due to tape zero error is

$$Z = -Csr_0 \tag{7.20}$$

where $C$ = the proportionality constant and is equal to (torque due to tape/torque due to precession) (refer *Section 7.2.2(2)*).

$s$ = value of one gyro scale unit in angular measure

It should be noted, that if the tape zero correction $Z$ is not applied to the observations for the instrument calibration constant $K$, then it will become part of $K$ and a further cause of variation in its value.


## (3)  Circle drift

This is movement of the horizontal circle caused possibly by vibration of the gyro. Thus the RO of the base line should be observed before and after the gyro observations, and the mean taken.


## (4)  Change of collimation and eccentricity of collimation

Difficult to eliminate as one cannot change face with some types of gyro-theodolite. Observational procedures can be adopted to reduce this error.


## (5)  Circle eccentricity

This form of error has already been discussed in Chapter 4 and can be reduced in gyro work by rotating the circle through 180° relative to the RO between sets of observation.


## 7.2.4 Convergence of meridians

Since gyroscopes establish true north, the underground base line bearing must be corrected before it can be related to a local or national grid. The correction $\delta$ can be illustrated from *Figure 7.11* on which the surface base $AB$ has a grid bearing due east, the origin of the grid being a point 0 on the Greenwich meridian. Assuming the underground base is parallel to the surface base, its bearing if fixed gyroscopically from $C$ will be $(90° - \delta)$, the direction of true north being $CN$. It can be seen from the illustration that the error $\delta$ is due to the convergence of meridians.

If the engineering project is based on the British national grid, the correction $\delta$ can be calculated using the OS Projection Tables for the Transverse Mercator Projection (TMP) of GB. At $(t - T)$ correction would be required only on very long lines, and therefore need not be considered for underground work. A similar correction would also be necessary if the survey was of small extent and based on a local grid (refer to *Engineering Surveying*, Volume 2, Chapter 2).

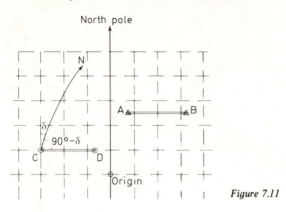

Figure 7.11

## 7.2.5 Observation procedures

Examples of the two basic methods of observation with a gyro-theodolite will now be given to illustrate more clearly the theory already covered.

## (1)  Reversal point method

The spin axis is brought approximately to north, the rotor accelerated to full speed and carefully suspended. As the gyro, as indicated on the gyro scale, starts to move away from the centre-line of the scale it is followed and kept on the centre-line of the scale, by rotating the tangent screws of the theodolite. As it reaches, say, its left reversal point ($r_1$, *Figure 7.8(a)*) movement ceases for a few seconds and the theodolite horizontal circle is read. The gyro is then tracked back, keeping it on the zero of the gyro scale, to its right reversal point $r_2$ and the theodolite again read. Thus the movement of the gyro is followed by simply keeping it on the centre-line of the gyro index, whilst the amplitude of its movement is measured by the theodolite.

| Reversal | Horiz circle reading |
|----------|----------------------|
| $r_1$ (left) | 42° 00′ 31″ |
| $r_2$ (right) | 49° 40′ 32″ |
| $r_3$ (left) | 42° 04′ 02″ |
| $r_4$ (right) | 49° 37′ 21″ |

Schuler's mean $N_1 = \dfrac{1}{4}(r_1 + 2r_2 + r_3)$

$$= \frac{1}{4}(42°\ 00'\ 31'' + 99°\ 21'\ 04'' + 42°\ 04'\ 02'')$$

$$= 45°\ 51'\ 24''$$

$$N_2 = \frac{1}{4}(r_2 + 2r_3 + r_4) = 45°\ 51'\ 29''$$

$$\therefore\ N = (N_1 + N_2)/2 = 45°\ 51'\ 26''\ \text{(horizontal circle reading of gyro north)}.$$

Assuming a tape zero correction $(Z)$ of $-10''$, then

$N_0 = N - Z = 45° 51' 26'' - 10'' = 45° 51' 16''$

A check on the computation can be applied by combining $N_1$ and $N_2$ to give

$$N = \frac{1}{8}(r_1 + 3r_2 + 3r_3 + r_4)$$

Assume that the theodolite is now sighted along the base line and the mean horizontal circle reading was, say, $55° 51' 16''$, then the base is obviously 10° clockwise of gyro north and its bearing relative to gyro north is therefore 10°.

The application of the instrument constant $K$ equal to, say, $-03' 00''$ will reduce the bearing of the base line to its correct geographical azimuth, i.e. $09° 57' 00''$ relative to true north.

If the survey is to be connected into the British national grid then a correction $(\delta)$ for convergence of meridians (i.e. the difference between grid north and true north) must be made and possibly a $(t - T)$ correction (i.e. the difference between an observed bearing and its corresponding grid bearing). In some instances a Laplace correction may need to be applied where the 'deviation of the vertical' is very high; however, in most cases this correction is usually less than $3''$ of arc. (Students should refer to Volume 2 of this work for details of the above corrections and worked examples.)

In the case of local surveys, of limited extent, on a rectangular Cartesian grid, only the convergence of meridians need be considered.

*Figure 7.12* shows the relationship of the various corrections.

Gyro bearing            $AB = \beta = $    $10° 00' 00''$
Instrument constant        $K = -0° 03' 00''$

Geographical azimuth      $\theta = $    $09° 57' 00''$
Convergence of meridians   $\delta = $    $+43' 09''$    computed from
$(t - T)$                          $= $    $+00' 04''$    geodetic tables

NG bearing of base-line    $AB = $   $10° 40' 13'' = \phi$

$(N.B.$ $Z$ is the tape zero correction $= -Csr_0.)$

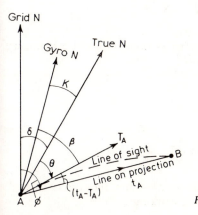

Grid N

Gyro N    True N

*Figure 7.12*

## (2) Transit method

In this method the gyroscope is set approximately to north ($N'$) and the whole instrument clamped. Oscillation of the gyro index about the gyro scale is noted and timed. For instance, with the gyro on the centre-line of the scale the reading is zero and the time is zero. As the gyro reaches its left reversal point ($A_W$) the gyro-scale reading is noted; as it returns to zero the time $t_2$ is noted (refer *Figure 7.9*). Similarly, the gyro-scale reading of the right reversal is noted ($A_E$) and the time $t_3$, when the gyro once again returns to the zero of the gyro index. The field data are reduced as follows:

| Transit time $t$ | Oscillation time $T$ | Time difference '$\Delta T$' | Reversal readings $A_W/A_E$ | $\frac{1}{2}(A_W + A_E)$ '$A$' | Horizontal circle $N'$ | $\Delta N$ |
|---|---|---|---|---|---|---|
| 0 m 00.0 s | | | | | 45° 47′ 00″ | |
| | −3 m 16.1 s | | −11.8 ($A_W$) | | | |
| 3 m 16.1 s | | +7.2 s | | 12.35 | | +4.25′ |
| | +3 m 23.3 s | | +12.9 ($A_E$) | | | |
| 6 m 39.4 s | | +7.7 s | | 12.35 | | +4.55′ |
| | −3 m 15.6 s | | −11.8 | | | |
| 9 m 55.0 s | | +7.6 s | | 12.35 | | +4.49′ |
| | +3 m 23.2 s | | +12.9 | | | |
| 13 m 18.2 s | | | | | *Mean* | +4.43′ |

$$\Delta N = CA\Delta T$$

where   $C$ = proportionality constant = 0.047 8 min of arc/div of gyro scale × sec of time

$$A = \frac{1}{2}(A_W + A_E) = \frac{1}{2}(11.8 + 12.9) = 12.35$$

$\Delta T$ = algebraic sum of times = $(-3 \text{ m } 16.1 \text{ s} + 3 \text{ m } 23.3 \text{ s})$
= +7.2 s ($T_W$ is −ve, $T_E$ is +ve)

then   $\Delta N = 0.0478 \times 12.35 \times 7.2 = 4.25'$

Horizontal circle reading of gyro $N = N' + \Delta N = 45° 51' 26''$

which is now corrected for $Z$ and/or $K$, as previously.

The value of $C$ is easily obtained using the above method with the instrument oriented say west of north, then east of north. Thus

$$N = N'_W + CA_1\Delta T_1 = N'_E + CA_2\Delta T_2$$

from which

$$C = \frac{N'_W - N'_E}{A_2\Delta T_2 - A_1\Delta T_1}$$

The main point to be emphasized about gyro observations is that they can be carried out on a line situated anywhere in the underground workings.

## 7.3 LINE AND LEVEL

### 7.3.1 Line

The line of the tunnel, having been established by wire survey or gyro observations, must be fixed in physical form in the tunnel. For instance, in the case of a Weisbach triangle (*Figure 7.13*) the bearing $W_u W_1$ can be computed; then, knowing the design bearing of the tunnel, the angle $\theta$ can be computed and turned off to give the design bearing, offset from the true by the distance $X W_u$. This latter distance is easily obtained from right-angled triangle $W_1 X W_u$.

The line is then physically established by carefully lining in three plugs in the roof from which weighted strings may be suspended as shown in *Figure 7.14(a)*. The third string serves to check the other two. These strings may be advanced by eye for short distances but must always be checked by theodolite as soon as possible.

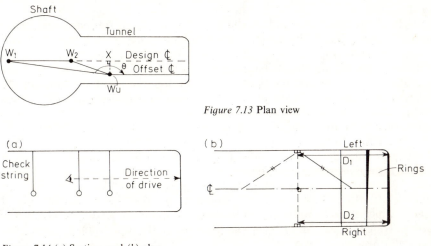

*Figure 7.13* Plan view

*Figure 7.14* (a) Section, and (b) plan

The gradient of the tunnel may be controlled by inverted boning rods suspended from the roof and established by normal levelling techniques.

Where tunnel shields are used for the drivage, laser guidance systems may be used for controlling the position and attitude of the shield. A laser beam is established parallel to the axis of the tunnel (i.e. on bearing and gradient) whilst a position-sensing system is mounted on the shield. This latter device contains the electro-optical elements which sense the position and attitude of the shield relative to the laser datum. Immunity to vibrations is achieved by taking 300 readings per second and displaying the average. Near the sensing unit is a monitor which displays the displacements in mm automatically corrected for roll. Additionally, roll, lead and look-up are displayed on push-button command along with details of the shield's position projected 5 m ahead. When the shield is precisely on line a green light glows in the centre of the screen. All the above data can be relayed to an engineers' unit several hundred metres away. Automatic print-out of all the data at a given shield position is also available to the engineers. The system briefly described here is the TG-26 system, designed and manufactured by ZED Instruments Ltd, Twickenham.

In addition to the above, 'square marks' are fixed in the tunnel by taping equilateral triangles from the centre-line or, where dimensions of the tunnel permit, turning off 90° with a theodolite, *Figure 7.14(b)*. Measurements from these marks enable the amount of lead in the rings to be detected. For instance, if $D_1 > D_2$ then the difference is the amount of left-hand lead of the rings. The gap between the rings is termed *creep*. In the vertical plane, if the top of the ring is ahead of the bottom this is termed *overhang*, the reverse is *look-up*. All this information is necessary to reduce to a minimum the amount of 'wriggle' in the tunnel alignment.

## Laser installation

In general the power output of commercial lasers is of the order of 5 mW and the intensity at the centre of a 2-cm diameter beam approximately 13 mW/cm². This may be compared with the intensity of sunlight received in the tropics at noon on a clear day, i.e. 100 mW/cm². Thus, as with the Sun, protective goggles should be worn when viewing the laser.

Practically all the lasers used in tunnelling work are wall- or roof-mounted, hence their setting is very critical. This is achieved by drilling a circular hole in each of two pieces of plate material, which are then fixed precisely on the tunnel line by conventional theodolite alignment. The laser is then mounted a few metres behind the first hole and adjusted so that the beam passes through the two holes and thereby establishes the tunnel line. Adjustment of the holes relative to each other in the vertical plane would then serve to establish the grade line.

An advantage of the above system is that the beam will be obscured should either the plates or laser move. In this event the surveyor/engineer will need to 'repair' the line, and to facilitate this, check marks should be established in the tunnel from which appropriate measurements can be taken.

In order to avoid excessive refraction, when installing the laser the beam should not graze the wall. Earth curvature and refraction limit the laser line to a maximum of 600 m, after which it needs to be moved forward. To minimize alignment errors, the hole farthest from the laser should be about one third of the total beam distance from the laser (refer *Sections 8.7.2 et seq.*).

### 7.3.2 Level

In addition to transferring bearing down the shaft, the correlation of the surface level to underground must also be made.

One particular method is to measure down the shaft using a 30-m standardized steel band. The zero of the tape is correlated to the surface BM as shown in *Figure 7.15*, and the other end of the tape precisely located using a bracket fixed to the side of the shaft. This process is continued down the shaft until a level reading can be obtained on the last tape length at B. The standard tension is applied to the tape and tape temperature recorded for each bay. A further correction is made for elongation of the tape under its own weight using:

Elongation (m) = $WL/2AE$

where    $E$ = modulus of elasticity of steel (N/mm²)
         $L$ = length of tape (m)
         $A$ = cross-sectional area of tape (mm²)
         $W$ = mass of tape (N)

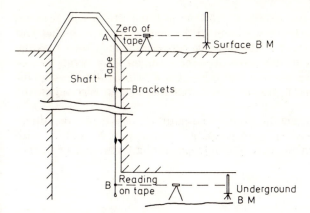

*Figure 7.15*

Then the corrected distance *AB* is used to establish the value of the underground **BM** relative to the surface **BM**.

If a special shaft tape (1000 m long) is available the operation may be carried out in one step.

The operation should be carried out at least twice and the mean value accepted. Using the 30-m band, accuracies of 1 in 5000 are possible, the shaft tape gives accuracies of 1 in 10 000.

Electromagnetic distance-measuring (EDM) equipment has also been used to measure shaft depths. A special reflecting mirror at the top of the shaft is aligned with the EDM instrument and then rotated in the vertical plane until the measuring beam strikes a reflector at the shaft bottom. In this way the distance from the instrument to the reflector is obtained and subsequently adjusted to give the distance from mirror to reflector. By connecting the mirror and reflector to surface and underground BM respectively, their values can be correlated. With top-mounted EDM a reflecting mirror is unnecessary and the distance to a reflector at the shaft bottom could be measured direct.

## 7.4 HYDROGRAPHIC SURVEYING

Hydrographic surveys are carried out in connection with harbour and dock construction, coastal defence work, sewage disposal, etc. Therefore, the engineer who works on such sites requires a knowledge of tide and wave theory in addition to basic surveying techniques.

### 7.4.1 Tidal theory

Both Newton and Laplace investigated tidal activity, but neither of their theories can account for the many variables such as irregular land masses, differing depths of water, etc., which are involved.

The main tide-producing forces are the gravitational attractions of the Moon and, to a lesser extent, the Sun, the ratio being 2.34:1.

Considering a particle of water on the Earth's surface, the Moon will exert an attraction on this particle directly proportional to the mass of the two bodies and inversely proportional to the square of the distance between them. However, as the Earth itself will also be subject to this attraction, the resultant force acting on the particle is the difference between the two and is called a *tide generating force*. This differential attraction is very small, $9 \times 10^{-7}$, and therefore has no effect upon the Earth's crust.

Considering the Moon's attraction on the Equator (*Figure 7.16*), the direct attractions at $E_1$ and $E_2$ will produce corresponding depressions at $N_1$ and $N_2$, with intermediate directions between the two. The tide-generating forces can be further

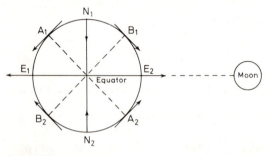

*Figure 7.16*

resolved into two components, the horizontal one being called the *tractive force*. These tractive forces serve to move the water from $N_1$ and $N_2$ towards $E_1$ and $E_2$, having their greatest intensity at points $A_1 A_2 B_1 B_2$. In this way high tides are formed at $E_1$ and $E_2$, and low tides at $N_1$ and $N_2$.

In practice, tidal predictions are based on the harmonic analysis of past records, commencing with an analysis of tidal curves obtained from self-registering tide gauges. These data are then used to prepare a predicting machine for future tidal information.

### 7.4.2 Tidal nomenclature

(1) *Spring tides* are the highest of the month and occur when the joint effect of the Sun's and Moon's attractions is at maximum (new or full Moon). In practice, the tide occurs some small interval (1 to 2 days) after the theoretical time; this is called the *age of the tide*.

(2) *Equinoctial spring tides* are exceptionally high, occurring during the equinoxes when the Sun and Moon are vertically over the Equator.

(3) *Neap tides* are the lowest of the month, occurring when the Sun's and Moon's attractions are in opposition to each other.

The average interval between corresponding tides on successive days is 24 h 50 min, thus each tide occurs 50 min later each day.

(4) *Lowest astronomical tide* is the sounding datum used for Admiralty charts and is the level of the lowest predicted tide.

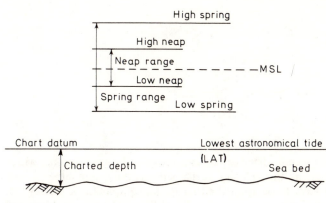

Figure 7.17

## 7.4.3 Survey techniques

Broadly the work may be divided into *on-shore* and *off-shore*. The former is carried out by the usual processes of triangulation, traversing, tacheometry, etc., while off-shore may be categorized as follows:

(A)  Vertical measurement of depths by sounding.
(B)  Horizontal control of the sounding position.
(C)  Reduction of soundings to a datum (Ordnance or tidal datum).

## (A)  Soundings

This is the practice of measuring the vertical depth from the water surface to the bed level.

In shallow depths of 5 m or less, a graduated wooden pole may be used. The pole is sometimes fitted with a cup to prevent it from sticking in the bed material. The method is slow and in a moving boat requires a certain skill.

In greater depths a sounding line is used. This is a wire, chain or hemp line with a cup-shaped lead weight attached. To sound a vertical depth, it is thrown ahead of the boat, so that the weight touches bottom when the boat is directly overhead. The weights vary from 2 to 5 kg depending upon the depth of water. The maximum length of line is 60 m, though for harbour soundings 20 m would suffice.

The personnel required for such work would consist of boat crew, leadsman, booker and, if the position was being fixed by three-point resection, a further two on sextants. The data booked would be number, depth and time of sounding, plus sextant angles if necessary.

For large areas requiring a continuous profile, echo-sounding apparatus would be used. This apparatus consists essentially of a transmitter, a receiver and a recorder. A sound pulse transmitted from the bottom of the ship at $T$ (*Figure 7.18(a)*), is reflected at $A$ and received at $R$, the time of travel being recorded. As the velocity of sound through water is known then the depth can be found. In practice, the returning echoes are picked up on an oscillator and converted from sound waves to high-frequency oscillations. These oscillations are amplified and transformed into a suitable current for operating a stylus on a travelling roll of paper.

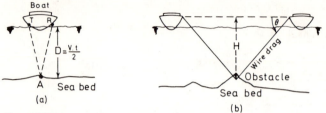

Figure 7.18

The apparatus is generally standardized for a mean velocity of sound through water equal to 1500 m/s. However, as the velocity varies with salinity, temperature and pressure, a correction may have to be applied either to the equipment or the results. Slight error will also occur due to the separation of $T$ and $R$, and this may be corrected by Pythagoras.

On hard rock bottoms most of the pulse energy is reflected back and produces a well-defined trace. On muddy bottoms the energy is absorbed, and the trace less clear. In this way bed material may be recognized. The equipment is accurate to about 1 in 200.

The transducer of the echo-sounder is reversible and used for the transmission and reception of the sound waves. The sound waves may be directed in either a wide beam or a narrow beam, i.e. a 55° cone or 6° to 2° cone. The narrow beam gives greater accuracy and is less prone to reflections from other sources; it must, however, be stabilized at extra cost. The wider beam gives greater coverage of an area resulting in fewer sounding lines. Indeed, the diameter of the area covered by the cone of sound, which will increase with depth, controls the separation of the sounding lines. A disadvantage of the wide beam is that it measures minimum depth which, due to the large area covered, may not be the depth directly below the ship; hence, interpretation of the echogram may be difficult. In practice a 30° cone is the one most commonly used.

As stated, most echo-sounders are calibrated for an average velocity of sound through water at 1500 m/s, by a controlled speed of 3000 rev/min. However, as velocity varies with water temperature, salinity and density, the apparatus must be adjusted to meet local conditions. The methods employed, in order of popularity, are:

(1) *By direct calibration* using a bar or target set horizontally below the transducer at known depths. The echo-sounder recorder is then adjusted to record those depths.
(2) *By computation* of the local velocity using measurements of temperature and salinity. This local velocity may then be used to calculate the operational rev/min of the apparatus. Calibration curves to facilitate this procedure are frequently supplied with the echo-sounder.

It should be noted that neither method is entirely satisfactory.

Apparatus used in conjunction with the echo-sounder is the *transit sonar* which makes an acoustic sweep of lanes 200 m in width. As lines of echo soundings may be 30 to 100 m apart, the presence of underwater obstacles may easily be missed. Using the sonar, a beam is emitted at right-angles to the path of the vessel, echoes from which produce an acoustic picture of the sea bed showing the presence of obstacles and changes in bed texture. The information is purely qualitative, but can be used to facilitate the interpretation of underwater contours, the study of pipeline routes and the inspection of areas of proposed construction.

Sweeping is normally carried out using a wire drag suspended at a known depth, $H$, between two vessels 100 m apart and moving in parallel paths (*Figure 7.18(b)*). When an obstruction is touched, the boats stop and the angle $\theta$ is measured. In this way a fix is made on the obstruction.

## (B)  Horizontal control of the sounding position

### (1)  In-shore methods

(a) *Cross-rope method* is used where the river or channel is narrow enough to permit a rope to be slung across (maximum width 300 m). The boat is pulled to a measured position on the rope and soundings taken. The method is relatively accurate and is essential where high velocities are encountered, such as near a waterfall. It is important that the position of the cross-rope is correlated to existing surveys.

(b) *Tacheometry* may be used in placid water, the staff being held in the boat.

(c) *Range and shore angle method* is indicated in *Figures 7.19* and *7.20*. The ranges are steering lines defined by objects on the shore; if the boat is kept on this line, then one angle from a base line will serve to fix its position. The method in *Figure 7.20* will give denser soundings near the shore.

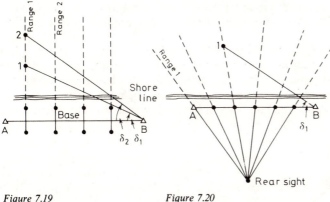

Figure 7.19                    Figure 7.20

(d) *Range and boat angle* is the reverse of the range and shore angle method, the angle being measured by sextant from the boat.

(e) *Simultaneous angles from the shore* as shown in *Figure 7.21* will establish position where no ranges exist. The boat is steered in an approximately straight line, the angles being measured on a pre-arranged signal or radio contact at the instant of sounding. An excellent check is obtained by using observations from three shore stations.

(f) *Simultaneous angles from the boat* requires two observers using sounding sextants to measure the angles to three known shore stations (*Figure 7.22*), and this is probably the most popular method.

The position of the boat may be fixed mechanically with the use of a *station pointer*. This is a circular protractor with three long straight-edge arms radiating from its centre. The two outer ones are movable so that the angles $\alpha$ and $\beta$ may be set. When the instrument is laid flat on the plan with the three arms passing through the three shore stations, the centre of the protractor gives the boat's position.

The same result may be achieved by constructing the three lines on tracing paper and applying the paper to the plan in the same way.

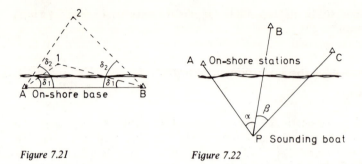

*Figure 7.21*                    *Figure 7.22*

In practice, however, the most popular method is by plotting *circle charts*, the principle of which is shown in *Figure 7.23*. The intersection of two circles passing through shore stations $A$, $B$ and $C$ gives the position of the boat at $P$. It can be clearly seen that the radius of circle $BC$ is $\frac{1}{2}BC$ cosec $\beta$; thus for various values of $\alpha$ and $\beta$ a chart can be produced of a large number of such circles having their respective centres on the perpendicular bisector of the chords $AB$ and $BC$. Interpolation between the circles for the measured angles $\alpha_n$, $\beta_n$, will fix the boat's position.

If by chance $P$ should fall on the circumference of a circle passing rhrough $A$, $B$ and $C$, no solution is possible. This is referred to as the *danger circle* and should be carefully avoided.

The range of the sextant is from 200 to 5000 m with an accuracy of about $\pm 4$ m.

(g) *Subtense methods* are widely used in harbour work. The horizontal method involves measuring the angle to both ends of a *fixed* horizontal distance whilst travelling along a range normal to the middle of the horizontal base; thus the angle will vary with distance from the base. The vertical method uses a vertical board marked in intervals

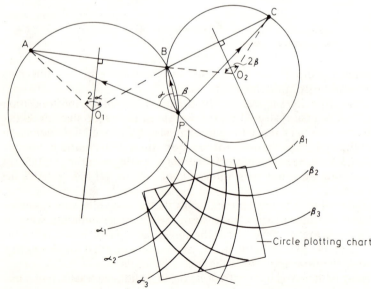

*Figure 7.23*

representing a known horizontal distance relative to a prefixed vertical angle. The zero of the board is established on the eye level of the sextant observer in the boat. With a fixed vertical angle on the sextant the observer views the zero mark, when the first interval mark comes into view on the sextant, the distance $D_1$ from the board is known and if the boat is kept on a fixed range in line with the board, its off-shore position is fixed. The method is quite accurate for the relatively short distances involved.

These methods constitute the optical systems commonly in use for what might be termed *in-shore* work.

In such work the distance apart of the lines of soundings will vary with the diameter of the cone of the echo-sounder and the depth of sounding. A system used by the Royal Navy specifies that on the scale of the plan:

(i) Sounding lines should not exceed 10 mm apart.
(ii) Fixes of the boat on the lines should not exceed 25 mm.

Using these specifications, combined with a knowledge of depth and angle of echo cone, the operation can be carefully planned and carried out.

## (2) *Off-shore methods* (*electromagnetic position fixing*)

(a) *Short range:* The equipment is portable, microwave equipment, comprising two on-shore 'remote' units which form a base line and give continuous distances to a receiver system on board ship. The ship's position is thus at the apex of a triangle whose three sides are known.

Two well-known systems are the Decca Trisponder 202A (range 80 km, accuracy ±3 m) and the Tellurometer Hydrodist MRB 201 (range 50 km, accuracy ±1.5 m). The systems operate at speeds up to 30 knots and can be linked to dynamic position-fixing systems capable of automatic operation and fitted with full data output facilities for the operation of computers, plotters, data recorders, etc.

(b) *Medium range: Figure 7.24* shows a 'master' station $A$, combined with two 'slaves' $B$ and $C$, which, in effect, generate a hyperbolic lattice of electromagnetic wave pattern over the area. With the aid of on-board phase meters, the ship's position within the lattice can be defined by hyperbolic co-ordinates. Correlation of the on-shore units into an appropriate survey system enables the hyperbolic co-ordinates to be converted to geographical or rectangular.

A well-known example of such a system is the Decca Hi-Fix 6 (range 300 km, accuracy $\sigma = 0.01$ lanes). The lane width along the base lines $AB$ and $AC$ is 75 m giving an accuracy of ±0.75 m, but this falls off rapidly as the lines diverge from each other. This fault can be eliminated by using two on-shore 'slaves' and a 'master' on board ship, resulting in a lattice of intersecting circles with a constant lane width. The disadvantage is that only one ship at a time may operate in this latter arrangement.

(c) *Satellite doppler:* Position-fixing on a world-wide basis, by the use of artificial satellites, became generally available in 1967 when the US Navy Navigational Satellite System-Transit, was brought into operation. At present the system has an absolute positioning accuracy of 1 to 2 m, and sub-metre accuracy in the relative positioning of points.

When a sound source and receiver are in motion relative to each other, the frequency of a transmitted wave at the receiver will differ from the frequency at the source according to a definite mathematical form. This is called the *Doppler effect* and is the principle used in satellite fixation. The orbiting satellites are tracked from accurately-

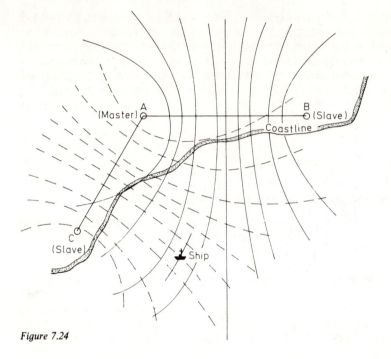

*Figure 7.24*

located ground stations and their position continuously updated. Thus the satellite becomes a navigational beacon which regularly transmits information defining its position. A vessel possessing the appropriate equipment receives the transmitted data, measures the doppler shift and converts the information into geographical co-ordinates defining the ship's position.

For more details of all these systems, reference should be made to specialist texts.

(d) *Acoustic techniques:* These particular techniques are useful for the precise position-fixing necessary in coring surveys, rig positioning, pipeline work, dynamic position of vessels, and the setting out of underwater construction.

Having established the position of the site using the aforementioned techniques, one may then require the equivalent of an accurate setting-out grid correctly positioned within the site area. This is done by putting down a pattern of sea-bed acoustic transformers (*Figure 7.25*) which can be positioned relative to each other to an accuracy of about 1 in 1000. The absolute position of the transponders is fixed by the methods already outlined, i.e. Decca Hi-Fix (only daytime operation possible), Aqua-Fix or Sat Nav. Several passes over the transponders will be made and their positions accurately assessed from all available data.

To position a vessel within the pattern area, the transponders are interrogated as shown. The time taken for the vessel to receive the return pulse from a particular beacon is a measure of the range. Position can be fixed to an accuracy of ±4.5 m from a minimum of two ranges. For stationary situations, such as fixing the position of a drilling vessel, repeated observations can be made to give an improved accuracy of about ±2.5 m.

The pattern of acoustic transponders can be varied according to requirements. For instance, a single transponder placed over an old drilling head is all that is required to

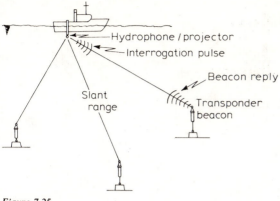

*Figure 7.25*

position a vessel directly over it. On the other hand, a series of transponders along the route of a pipeline would be necessary for its location.

The advantages of the acoustic system are:

(i) No shore-based transmitting sites are required once the transponder positions have been fixed.
(ii) The system can be used continuously.
(iii) The system can be used anywhere and, once established, it is available anytime, for it can be left on location indefinitely and survey can be resumed at any time without loss of accuracy.

## (C) Reduction of soundings

In order to reduce soundings, it is necessary to know the level of the water surface, relative to some datum, at the instant of sounding. To achieve this a *tide gauge* is used.

In its simplest form this gauge consists of a graduated post fixed either on the shore, or on the side of a quay wall. Such a gauge is called a *staff gauge* (*Figure 7.26(a)*) and is generally related to OD by direct levelling.

Where the whole tidal range requires investigation, it may be necessary to establish a series of gauges from high-water level to low-water level. The same applies along the shore line where there is a variation in water level (*Figure 7.26(b)*).

Where a continuous record of the tides is required *self-registering gauges* are used (*Figure 7.27*). These are generally situated in tidal observatories throughout the

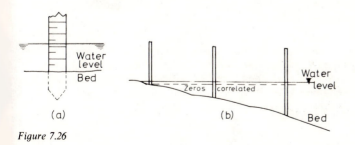

*Figure 7.26*

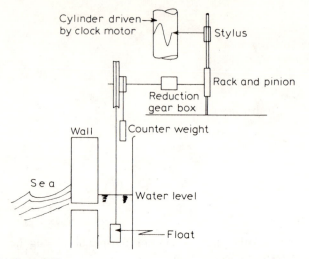

*Figure 7.27*

country and consist of a float hung in a well, with a pipe outlet to the sea set well below low water level. The float is connected by a flexible wire, through a series of pulleys and gears, to a stylus. The change of water level on the float is recorded by a fine line on a travelling roll of paper.

Alternative tide gauges may be used, such as the *float gauge* (*Figure 7.28*) in which the graduated staff is attached to a float and enclosed within a box. The box is perforated at its lower end to allow the sea access and thus acts as a stilling chamber. As the staff is graduated from zero at sea-level, the depth to sea-level is read off against an index mark which has a value relative to OD. Thus the level of the sea at any instant of time is known.

As staff gauges require constant observation, a self-registering gauge of the type already mentioned is generally used. However, typical of modern developments in this area, a recording tide gauge capable of operating on the sea-bed at depths of 200 m in

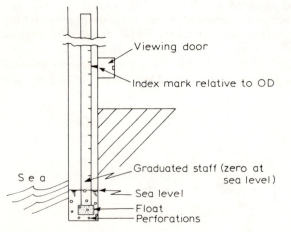

*Figure 7.28*

remote locations, has been developed at the National Institute of Oceanography. The instrument consists of a 560-mm diameter sphere which houses electronic systems and batteries. This is mounted in a tubular tripod that sits on the sea bed. External to the sphere are a pressure transducer, temperature sensor and acoustic command and transmit beacons. On command from the surface the sphere assembly floats to the surface for retrieval. The basic principal of the equipment is based on the measurement of flexure of a diaphragm under varying hydrostatic pressure.

If tide prediction over a local area is required, tide gauge readings should be taken over a minimum period of two weeks so as to include spring and neap tides. The readings may be at hourly intervals, increasing to 5-minute intervals about the time of high or low water. For the reduction of soundings they should be at 5 to 10-minute intervals over the period of sounding.

As the variation of the water level is not linear, a graph should be plotted of tide gauge reading against time. In this way values can be interpolated for the instant of sounding as follows. Assume the zero of the tide gauge is 1 m OD (obtained by direct levelling), and tide gauge readings at 9.00 a.m. and 9.10 a.m. were 1.08 and 1.1 m respectively. At sea a sounding of 10 m was recorded for point $A$ at 9.05 a.m.

By interpolation from the graph the reading at 9.05 a.m. is 1.09 m.

Therefore the water level, at this instant, relative to OD $= 1 + 1.09 = 2.09$ m OD. As the sounding was 10 m, the level of point $A$ would be $(2.09 - 10) = -7.91$ m OD.

To eliminate the use of negative signs, it would be advisable to assume a value of 100 m for mean sea level, in which case the level of $A = 92.09$ m.

### 7.4.4 Sextant

A sounding sextant (*Figure 7.29*), as used in position fixing, is a more robust version of an ordinary sextant and has a greater sighting range. Its principles are as follows:

To measure the angle $AOB$, the target at $A$ is viewed direct through the unsilvered glass at $E$. The image of target $B$ is brought into coincidence with $A$ in the silvered portion of the mirror at $E$, by manipulating the mirror arm $CF$ to position $CG$. Now:

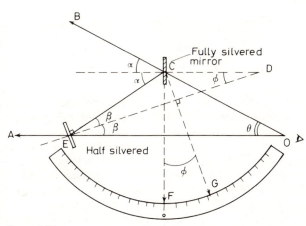

*Figure 7.29*

angle $BOA = 2\alpha - 2\beta = \theta$

angle $CDE = FCG = (\alpha - \beta) = \phi$

$$\therefore \ \theta = 2\phi$$

The scale of the sextant is so graduated that the double angle $\phi = \theta$, can be read direct. Thus the 75° arc permits angles up to 150° to be resolved to 1 minute of arc.

It should be noted that the angle $AOB$ is measured in the plane of the objects viewed.

Thus if $A$ and $B$ are at greatly differing elevations the observed angle would need to be reduced to the horizontal using the spherical trig equation (*Figure 7.30*)

$$\cos \theta_H = \frac{\cos \theta - \cos \alpha \cos \beta}{\sin \alpha \sin \beta} \tag{7.21}$$

where $\theta_H$ is the horizontal equivalent of the measured angle $\theta$, and $\alpha$, $\beta$ are vertical angles measured from the vertical, i.e.

$$\alpha = (90° \pm \delta_A) \quad \text{and} \quad \beta = (90° \pm \delta_B)$$

This can be avoided by keeping the terrain stations at shore level.

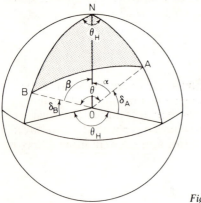

*Figure 7.30*

## 7.4.5 Direction and velocity of currents

This information is very necessary to the engineer, particularly in the case of sewage disposal. The simplest solution is to note the position and time of a float as it is carried through the water (*Figure 7.31*), although the surface does not behave in the same way as the main body of water. This is due to wind, the effect of cross-currents, and the fact that fresh-water streams entering sea water do not readily mix, and travel uppermost at a different velocity for some distance. Also, in any body of water the maximum velocity occurs some distance below the surface. In view of these considerations, the floats used should be so constructed that they extend well below the surface, with just the indicator at the surface. The position of the float can be fixed by any of the methods already outlined.

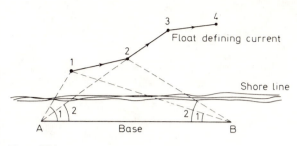

*Figure 7.31*

## 7.4.6 Engineering applications

Apart from the more obvious applications of pipeline operations, rig positioning and coring surveys in which the electromagnetic systems are continually used, the following serve to illustrate their smaller-scale use:

(1) *Trailer suction dredging* in which position-fixing is very critical—used in conjunction with track plotter to avoid over-dredging (low spots) or under-dredging (high spots). Receiver sited directly over suction head.

(2) *Precision dredging by grab barge:* It is frequently necessary to remove hard spots left by suction methods. This requires precision survey to locate the spots, and then precision positioning to remove them, finally post dredging surveys to confirm the work.

(3) *Obstacle and wreck sweeping:* In busy waterways in particular, the location and pin-pointing of obstacles is vitally important, as it would be to many engineering projects. Using an echo-sounder with a particularly wide cone and Hi-Fix track plotter to obtain lines about 5 m apart, it is possible to ensure that no gaps are left. If the search is unsuccessful, it is possible to 'interline' the previously searched lines using the track plotter, thus giving lines 2.5 m apart.

(4) *Float tracking* to determine the rate and direction of currents. This information may be required at certain depths below the surface, in which case 'kites' are suspended at the requisite depth below the buoy floats. For shallow depths, a pole of uniform cross-section is made to float vertically, with only sufficient of its upper end visible to allow observation. One method of tracking the floats is to utilize a Hi-Fix receiver in a small launch, with the receiving antenna mounted on an overside boom. A pointer, vertically beneath the antenna, enables it to be positioned directly over the buoy.

(5) *Isotope tracking* is used to investigate the subsequent movement of deposited dredged spoil, and that of effluent from sewer outfalls.

Ground glass is prepared to the grain size of the dumped spoil and exposed to radio-active isotopes.

Detection of the radio-active trace is made by towing a scintillation counter over the sea-bed. This is also done prior to dumping to correct for any existing radio-active count in the area. Radio-active tracer distribution contour charts are prepared from scintillation count surveys using a position-fixing system like Hi-Fix.

The radio-active life may be short or long, depending upon the period of investigation envisaged.

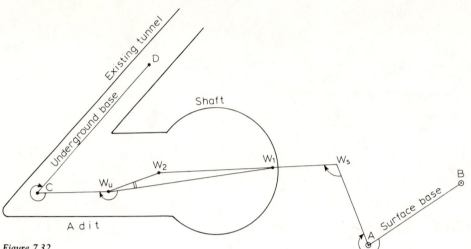

*Figure 7.32*

## Worked examples

*Example 7.1.* The national grid (NG) bearing of an underground base-line, *CD* (*Figure 7.32*), is established by co-planning at the surface on to two wires, $W_1$ and $W_2$, hanging in a vertical shaft, and then using a Weisbach triangle underground.

The measured field data is as follows:

| | |
|---|---|
| NG bearing *AB* | 74° 28′ 34″ |
| NG co-ords of *A* | E 304625 m, N 511612 m |
| Horizontal angles: | |
| $BAW_s$ | 284° 32′ 12″ |
| $AW_sW_2$ | 102° 16′ 18″ |
| $W_2W_uW_1$ | 0° 03′ 54″ |
| $W_1W_uC$ | 187° 51′ 50″ |
| $W_uCD$ | 291° 27′ 48″ |
| Horizontal distances: | |
| $W_1W_2$ | 3.625 m |
| $W_uW_2$ | 2.014 m |

In order to check the above correlation, a gyro-theodolite is set up at *C* and the following horizontal scale theodolite readings to the reversal points of the gyro recorded:

| | |
|---|---|
| Left reversal point | 336° 25′ 18″ |
| Right reversal point | 339° 58′ 52″ |
| Left reversal point | 336° 02′ 44″ |

The mean horizontal scale reading on to *CD* is 20° 51′ 26″.

| Given: | Convergence of meridians | ±0° 17′ 24″ |
|---|---|---|
| | $(t − T)$ correction | ±0° 00′ 06″ |
| | Instrument constant | −2° 28′ 10″ |

compute the difference between the bearings of the underground base as fixed by the wire survey and by the gyro-theodolite. (KP)

The first step is to calculate the bearing of the wire base using the measured angles at the surface:

Grid bearing of $AB$ = 74° 28′ 34″ (given)
Angle $BAW_s$ = 284° 32′ 12″

Grid bearing of $AW_s$ = 359° 00′ 46″
Angle $AW_sW_2$ = 102° 16′ 18″

Sum = 461° 17′ 04″
−180

Grid bearing $W_1W_2$ = 281° 17′ 04″ (refer *Figure 7.32*)

Now, using the Weisbach triangle calculate the bearing of the underground base from the wire base:

*From a solution of the Weisbach triangle*

$$\text{Angle } W_2W_1W_u = \frac{234'' \times 2.014}{3.625} = 130'' = 0° \, 02' \, 10''$$

Grid bearing $W_1W_2$ = 281° 17′ 04″
Angle $W_2W_1W_u$ = 0° 02′ 10″

Grid bearing $W_1W_u$ = 281° 14′ 54″
Angle $W_1W_uC$ = 187° 51′ 50″

Sum = 469° 06′ 44″
−180

Grid bearing $W_uC$ = 289° 06′ 44″
Angle $W_uCD$ = 291° 27′ 48″

Sum = 580° 34′ 32″
−540

Grid bearing $CD$ = 40° 34′ 32″ (underground base)

Compute the grid bearing of $CD$ using gyro data:

$$\text{Horizontal circle reading of gyro north} = \frac{1}{2}\left(\frac{r_1 + r_3}{2} + r_2\right)$$

$$= \frac{1}{2}[(336° \, 25' \, 18'' + 336° \, 02' \, 44'')/2 + 339° \, 58' \, 52''] = 338° \, 06' \, 26''$$

Horizontal circle reading of base $CD$ = 20° 51′ 26″
∴ Bearing of $CD$ relative to gyro north = 42° 45′ 00″
Bearing of $CD$ relative to true north = 42° 45′ 00″ − instr constant
= 40° 16′ 50″ = $\phi$ (*Figure 7.33*)

With reference to *Figure 7.33*, it can be seen how the 'convergence of meridians' ($\delta\theta$)

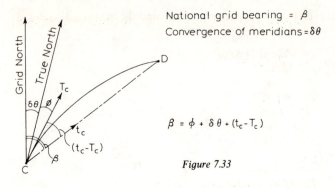

National grid bearing = $\beta$
Convergence of meridians = $\delta\theta$

$$\beta = \phi + \delta\theta + (t_c - T_c)$$

Figure 7.33

and $(t - T)$ are applied to transform true N to grid N (refer Volume 2, Chapter 2 for details).

$\therefore$ National grid bearing $CD = 40° 16' 50'' + 0° 17' 24'' + 0° 00' 06''$
$$= 40° 34' 20''$$

Difference in bearings $= 12''$.

*Example 7.2.* The centre-line of the tunnel *AB* shown in *Figure 7.34* is to be set out to a given bearing. A short section of the main tunnel has been constructed along the approximate line and access is gained to it by means of an adit connected to a shaft. Two wires *C* and *D*, are plumbed down the shaft, and readings are taken on to them by a theodolite set up at station *E* slightly off the line *CD* produced. A point *F* is located in the tunnel, and a sighting is taken on to this from station *E*. Finally a further point *G* is located in the tunnel and the angle *EFG* measured.

From the survey initially carried out, the co-ordinates of *C* and *D* have been calculated and found to be E 375.78 m and N 1119.32 m, and E 375.37 m and N 1115.7 m respectively.

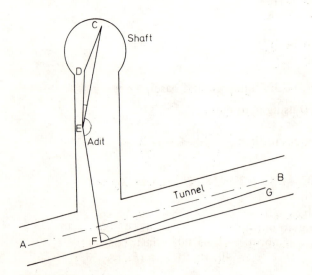

*Figure 7.34*

Calculate the co-ordinates of $F$ and $G$. Without making any further calculations describe how the required centre-line could then be set out. (ICE)

Given data: $CD =$ 3.64 m  $DE =$ 4.46 m
$EF =$ 13.12 m  $FG =$ 57.5 m
Angle $DEC =$ 38″
Angle $CEF =$ 167° 10′ 20″
Angle $EFG =$ 87° 23′ 41″

Solve Weisbach triangle for angle $ECD$

$$\hat{C} = \frac{ED}{DC}\,\hat{E} = \frac{4.46}{3.64} \times 38'' = 47''$$

*By co-ordinates*

Bearing of wire base $CD = \tan^{-1}\dfrac{-0.41}{-3.62} = 186° \; 27' \; 19''$

$\therefore$ WCB of $CE$    $= 186° \; 27' \; 42'' - 47'' = 186° \; 26' \; 55''$
WCB of $CE$    $= 186° \; 26' \; 55''$
Angle $CEF$    $= 167° \; 10' \; 20''$

WCB of $EF$    $= 173° \; 37' \; 15''$
Angle $EFG$    $= \;\; 87° \; 23' \; 41''$

WCB of $FG$    $= \;\; 81° \; 00' \; 56''$

| Line | Length (m) | WCB | Co-ordinates | | Total co-ordinates | | |
|------|-----------|-----|--------------|---|--------------------|---|---|
| | | | $\Delta E$ | $\Delta N$ | $E$ | $N$ | Station |
| | | | | | 375.78 | 1119.32 | C |
| CE | 8.10 | 186° 26′ 55″ | −0.91 | − 8.05 | 374.87 | 1111.27 | E |
| EF | 13.12 | 173° 37′ 15″ | 1.46 | −13.04 | 376.33 | 1098.23 | F |
| FG | 57.50 | 81° 00′ 56″ | 56.79 | 8.99 | 433.12 | 1107.22 | G |

Several methods could be employed to set out the centre-line; however, since bearing rather than co-ordinate position is critical, the following approach would probably give the best results.

Set up at $G$, the bearing of $GF$ being known, the necessary angle can be turned off from $GF$ to give the centre-line. This is obviously not on centre but is the correct line; centre positions can now be fixed at any position by offsets.

*Example 7.3.* Two vertical wires $A$ and $B$ hang in a shaft, the bearing of $AB$ being 55° 10′ 30″ (*Figure 7.35*). A theodolite at $C$, to the right of the line $AB$ produced, measured the angle $ACB$ as 20′ 25″. The distances $AC$ and $BC$ were 6.4782 m and 3.2998 m respectively.

Calculate the perpendicular distance from $C$ to $AB$ produced, the bearing of $CA$ and the angle to set off from $BC$ to establish $CP$ parallel to $AB$ produced.

Describe how you would transfer a line $AB$ above ground to the bottom of a shaft.

(LU)

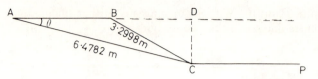

*Figure 7.35*

$AB \approx AC - BC = 3.1784$ m

$$\text{Angle } BAC = \frac{3.2998}{3.1784} \times 1225'' = 1272'' = 21'\,12'' = \theta$$

$$\text{By radians } CD = AC \times \theta \text{ rad} = \frac{6.4782 \times 1272}{206\,265} = 0.0399 \text{ m}$$

Bearing $AB = 55°\,10'\,30''$
Angle $BAC = \phantom{55°\,}21'\,12''$
_____

Bearing $AC = 55°\,31'\,42''$
∴ Bearing $CA = 235°\,31'\,42''$
Angle to be set off from $BC = ABC = 180° - (21'\,12'' + 20'\,25'')$
$\phantom{Angle to be set off from BC = ABC} = 179°\,18'\,23''$

*Example 7.4.* In order to survey the bed of a channel, soundings were taken at 30-m intervals on a square grid system, during the incoming tide, the boat proceeding in the directions *ABCD–K* and the soundings obtained were as shown below. The time when starting at *A* was 10 a.m. and when finishing at *K* 11.36 a.m. At these times the tide gauge readings were 6 and 12 m respectively.

If the zero of the gauge was 2 m OD find the reduced levels of the channel at the 25 sounding points, assuming uniform rate of rise of water level and uniform rate of operation from *A* to *K*.

| *A* | 3.0 | 3.2 | 3.3 | 3.5 | 3.6 | *B* |
|---|---|---|---|---|---|---|
| *D* | 7.7 | 7.3 | 7.0 | 6.6 | 6.3 | *C* |
| *E* | 8.5 | 8.7 | 8.8 | 9.0 | 9.1 | *F* |
| *H* | 10.1 | 9.7 | 9.3 | 9.0 | 8.7 | *G* |
| *J* | 7.8 | 8.0 | 8.1 | 8.3 | 8.4 | *K* |

Write brief notes on the equipment required and the methods of operation. If the method here is open to criticism suggest improvements.　(LU)

Interval between first and last soundings $= 96$ min
∴ Interval between each sounding $\phantom{xx}= 96 \div 24 = 4$ min
Tidal range during soundings period $\phantom{x}= 6$ m

$$\therefore \text{ Rise per 4-minute interval} \phantom{xxxx} = \frac{6 \times 4}{96} = 0.25 \text{ m}$$

Consider sounding *A*
Level of water surface at 10 a.m. $= 6 + 2 = 8$ m OD
Depth of sounding $\phantom{xxxxxxxxx}= 3$
∴ Reduced level of channel $\phantom{xxx}= 8 - 3 = 5$ m OD

The remaining levels are completed in the same manner:

| Station | Tide gauge reading | Level of water surface | Depth of sounding | RL of channel (m OD) |
|---|---|---|---|---|
| A | 6.0 | 8.0 | 3.0 | 5.00 |
| 1 | 6.25 | 8.25 | 3.2 | 5.05 |
| 2 | 6.50 | 8.50 | 3.3 | 5.20 |
| 3 | 6.75 | 8.75 | 3.5 | 5.25 |
| B | 7.00 | 9.00 | 3.6 | 5.40 |
| C | 7.25 | 9.25 | 6.3 | 2.95 |
| 1 | 7.50 | 9.50 | 6.6 | 2.90 |
| 2 | 7.75 | 9.75 | 7.0 | 2.75 |
| 3 | 8.00 | 10.00 | 7.3 | 2.70 |
| D | 8.25 | 19.25 | 7.7 | 2.55 |
| E | 8.50 | 10.50 | 8.5 | 2.00 |
| ⋮ | ⋮ | ⋮ | ⋮ | ⋮ |
| K | 12.00 | 14.00 | 8.4 | 5.60 |

*Criticism* (1) For the best results, the work should be carried out during slack water at high or low tide. During incoming tide the conditions are much less stable.

(2) Rate of rise of tide is not uniform; more tide-gauge readings required.

## Exercises

(*7.1*) (a) Describe fully the surveying operations which have to be undertaken in transferring a given surface alignment down a shaft in order to align the construction work of a new tunnel.

(b) A method involving the use of the three-point resection is often employed in fixing the position of the boat during off-shore sounding work.

Describe in detail the survey work involved when this method is used and discuss any precautions which should be observed in order to ensure that the required positions are accurately fixed. (ICE)

(*7.2*) Describe how you would transfer a surface bearing down a shaft and set out a line underground in the same direction.

Two plumb lines A and B in a shaft are 8.24 m apart and it is required to extend the bearing AB along a tunnel. A theodolite can only be set up at C 19.75 m from B and a few millimetres off the line AB produced.

If the angle BCA is 09′ 54″ what is the offset distance of C from AB produced? (ICE)

(*Answer:* 195 mm)

(*7.3*) Soundings were taken from a boat P whilst observations were made by sextant onto three shore signals A, B and C having co-ordinates (0, 0), (0, 850) and (325, 1375) m respectively. For one of the soundings the horizontal angles APB and BPC were 41° 30′, and 28° 20′ respectively, the boat being to the north-west of the area ABC.

Calculate the distance of the boat from station B, and check your calculation by a graphical construction. (ICE)

(*Answer:* 1220 m)

(*7.4*) The surface currents around a proposed sewer outfall into the sea are to be studied by plotting the drift of a float released at the appropriate times.

If it is practical to follow the float by a boat and remain in sight of several prominent features on shore, all identifiable on the 6″-scale Ordnance sheet of the area, how would you fix and plot the drift of the float? (ICE)

(7.5) In order to determine the cross-sectional profile of the bed of a tidal river, readings were taken from a theodolite fitted with stadia hairs onto a vertical staff held in a boat at the same time as readings of a sounding rod were observed in the boat and tide gauge readings were made on the shore. The base of the staff was not necessarily at the same level as the water surface. The reading on the tide gauge of 3.05 m has a reduced level of 4 m above datum.

The following observations were recorded:

| Point | Staff at | Stadia readings (m) | | | Vertical angle | Sounding rod (m) | Tide gauge (m) |
|-------|----------|-------|--------|-------|---------|---------|---------|
| | | upper | middle | lower | | | |
| 1 | HWM | 2.844 | 2.761 | 2.679 | $-5°$ | – | – |
| 2 | Boat | 2.094 | 1.957 | 1.820 | $-5°$ | 2.08 | 2.50 |
| 3 | Boat | 1.375 | 1.189 | 1.003 | $-5°$ | 3.56 | 2.46 |
| 4 | Boat | 0.744 | 0.503 | 0.262 | $-5°$ | 3.79 | 2.42 |
| 5 | Boat | 2.804 | 2.527 | 2.249 | $-2°$ | 3.29 | 2.38 |
| 6 | Boat | 2.618 | 2.304 | 1.990 | $-2°$ | 2.22 | 2.35 |
| 7 | Boat | 2.393 | 2.033 | 1.673 | $-2°$ | 0.99 | 2.29 |
| 8 | HWM | 1.874 | 1.487 | 1.100 | $-2°$ | – | – |
| – | 3.05 m mark on tide gauge | 4.160 | 3.978 | 3.795 | $0°$ | – | – |

In the above table all the numbered points lie on a straight line across the river, points 1 and 8 being at high water mark. The theodolite was placed at one end of this line.

Calculate the reduced levels of the eight points on the bed of the river and plot the required cross-sectional profile on graph paper. (ICE)

(*Answer:* 1 = 3.75, 2 = 1.32, 3 = $-0.18$, 4 = $-0.46$, 5 = 0, 6 = 1.04, 7 = 2.21, 8 = 3.75)

*Hint:* Use reading '3.05 m mark on tide gauge' to find the RL of the centre of theodolite in order to find RL of point 1.

# 8
# Setting out (dimensional control)

In engineering the production of an accurate large-scale plan is usually the first step in the planning and design of a construction project. Thereafter the project, as designed on the plan, must be set out on the ground in the correct absolute and relative position and to its correct dimensions. Thus, surveys made in connection with a specific project should be planned with the setting-out process in mind and a system of three-dimensional control stations conveniently sited and adequate in number, should be provided to facilitate easy, economical setting out.

It is of prime importance that the establishment and referencing of survey control stations should be carried out at such places and in such a manner that they will survive the construction processes. This entails careful choice of the locations of the control stations and their construction relative to their importance and long- or short-term requirements. For instance, those stations required for the total duration of the project may be established in concrete or masonry pillars with metal plates or bolts set in on which is punched the station position. Less durable are stout wooden pegs set in concrete or driven directly into the ground. A system of numbering the stations is essential, and frequently pegs are painted different colours to denote the particular functions for which they are to be used.

## 8.1 PROTECTION AND REFERENCING

Most site operatives have little concept of the time, effort and expertise involved in establishing setting out pegs. For this reason the pegs are frequently treated with disdain and casually destroyed in the construction process. A typical example of this is the centre-line pegs for route location which are the first to be destroyed when earth-moving commences. It is important, therefore, that control stations and BM should be protected in some way (usually as shown in *Figure 8.1*) and site operatives, particularly earthwork personnel, impressed with the importance of maintaining this protection.

Where destruction of the pegs is inevitable, then referencing procedures should be adopted to re-locate their positions to the original accuracy of fixation. Various configurations of reference pegs are used and the one thing that they have in common is that they must be set well outside the area of construction and have some form of protection, as in *Figure 8.1*.

A commonly-used method of referencing is from four pegs ($A, B, C, D$) established

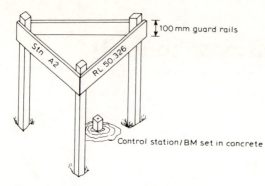

*Figure 8.1*

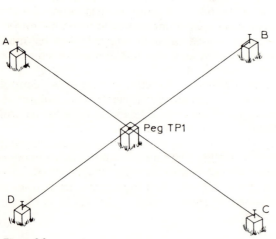

*Figure 8.2*

such that two strings stretched between them intersect to locate the required position (*Figure 8.2*). Distances *AB, BC, CD, AD, AC, BD* should all be measured as checks on the possible movement of the reference pegs, whilst distances from the reference pegs to the setting-out peg will afford a check on positioning. Intersecting lines of sight from theodolites at say *A* and *B* may be used where ground conditions make string lining difficult.

Where ground conditions preclude taping, the setting-out peg may be referenced by trisection from three reference pegs. The pegs should be established to form well-conditioned triangles of intersection (*Figure 8.3*), the angles being measured and set out on both faces of a 1″ theodolite.

All information relating to the referencing of a point should be recorded on a diagram of the layout involved.

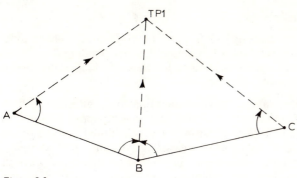

*Figure 8.3*

## 8.2 BASIC SETTING-OUT PROCEDURES USING CO-ORDINATES

Plans are generally produced on a plane rectangular co-ordinate system, hence salient points of the design may also be defined in terms of rectangular co-ordinates on the same system. For instance, the centre-line of a proposed road may be defined in terms of co-ordinates at, say, 30-m intervals, or alternatively, only the tangent and intersection points may be so defined. The basic methods of locating position when using co-ordinates is by either polar co-ordinates, or intersection.

### 8.2.1 By polar co-ordinates

In *Figure 8.4*, *A*, *B* and *C* are control stations whose co-ordinates are known. It is required to locate point *IP* whose design co-ordinates are also known. The computation involved is as follows:

(1) From co-ordinates compute the bearing *BA* (this bearing may already be known from the initial control survey computations).
(2) From co-ordinates compute the horizontal length and bearing of *B − IP*.
(3) From the two bearings compute the setting-out angle *AB(IP)*, i.e. *β*.
(4) Before proceeding into the field, draw a neat sketch of the situation showing all the setting-out data. Check the data from the plan or by independent computation.

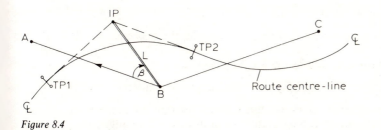

*Figure 8.4*

The field work involved is as follows:

(1) Set up theodolite at *B* and backsight to *A*, note the horizontal circle reading.
(2) Add the angle $\beta$ to the circle reading *BA* to obtain the circle reading $B - IP$. Set this reading on the theodolite to establish direction $B - IP$ and measure out the horizontal distance *L*.

If this distance is set out by steel tape careful considerations must be given to all the error sources such as standardization, slope, tension and possibly temperature if the setting-out tolerances are very small. It should also be carefully noted that the sign of the correction is reversed from that applied when measuring a distance. For example, if a 30-m tape was in fact 30.01 m long, when measuring a distance the recorded length would be 30 m for a single tape length, although the actual distance is 30.01 m; hence a POSITIVE correction of 10 mm is applied to the recorded measurement. However, if it is required to set out 30 m, the actual distance set out would be 30.01 m; thus this length would need to be reduced by 10 mm; i.e., a NEGATIVE correction.

The best field technique when using a steel tape is carefully to align pegs at *X* and *Y* each side of the expected position of *IP* (*Figure 8.5*). Now, as carefully, measure the

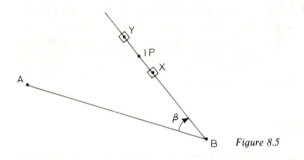

B       *Figure 8.5*

distance *BX* and subtract it from the known distance to obtain distance $X - IP$, which will be very small, possibly less than one metre. Stretch a fine cord between *X* and *Y* and measure $X - IP$ along this direction to fix point *IP*.

Modern EDM, such as the Aga Geodimeter 122, displays horizontal distance, so the length $B - IP$ may be ranged direct to a reflector fixed to a setting-out pole. The use of short-range EDM equipment has made this method of setting out very popular.

## 8.2.2 By intersection

This technique, illustrated in *Figure 8.6*, does not require linear measurements; hence, adverse ground conditions are immaterial and one does not have to consider tape corrections.

The computation involved is as follows:

(1) From the co-ordinates of *A*, *B* and *IP* compute the bearings *AB*, $A - IP$ and $B - IP$.
(2) From the bearings compute the angles $\alpha$ and $\beta$.

The relevant field work, assuming two theodolites are available, is as follows:

(1) Set up a theodolite at *A*, backsight to *B* and turn off the angle $\alpha$.
(2) Set up a theodolite at *B*, backsight to *A* and turn off the angle $\beta$.

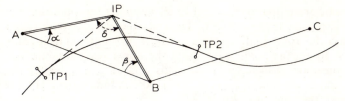

*Figure 8.6*

The intersection of the sight lines $A - IP$ and $B - IP$ locates the position of $IP$. The angle $\delta$ is measured as a check on the setting out.

If only one theodolite is available then two pegs per sight line are established, as in *Figure 8.5*, then string lines connecting each opposite pair of pegs locate position $IP$, as in *Figure 8.2*.

## 8.3 TECHNIQUE FOR SETTING OUT A DIRECTION

It can be seen that both the basic techniques of position fixing require the turning-off of a given angle. To do this efficiently the following approach is recommended:

In *Figure 8.6*, consider turning off the angle $\beta$ equal to $20°\ 36'\ 20''$ using a Watts No. 1 $(20'')$ theodolite (*Figure 8.7(a)*).

(1) With theodolite set at $B$, backsight to $A$ and read the horizontal circle—say, $02°\ 55'\ 20''$.
(2) As the angle $\beta$ is clockwise of $BA$ the required reading on the theodolite will be equal to $(02°\ 55'\ 20'' + 20°\ 36'\ 20'')$, i.e. $23°\ 31'\ 40''$.

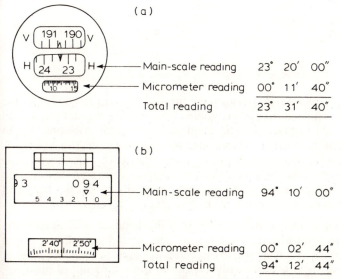

*Figure 8.7 (a)* Watts No. 1—20″ theodolite. *(b)* Wild T2—1″ theodolite

(3) As the *minimum* main scale division is equal to 20′ anything less than this will appear on the micrometer (*Figure 8.7(a)*). Thus, set the micrometer to read 11′ 40″, now release the upper plate clamp and rotate theodolite until it reads approximately 23° 20′ on the main scale; using the upper plate slow-motion screw, set the main scale to exactly 23° 20′. This process will not alter the micrometer scale and so the total reading is 23° 31′ 40″, and the instrument has been swung through the angle $\beta = 20° 36′ 20″$.

If the Wild T.2 (*Figure 8.7(b)*) had been used, an examination of the main scale shows its minimum division is equal to 10′. Thus, to set the reading to 23° 31′ 40″ one would set only 01′ 40″ on the micrometer first before rotating the instrument to read 23° 30′ on the main scale.

Therefore, when setting out directions with any make of theodolite, the observer should examine the reading system to find out its minimum main scale value, anything less than which is put on the micrometer first.

Basically the micrometer works as shown in *Figure 8.8*, and, if applied to the Watts theodolite, is explained as follows:

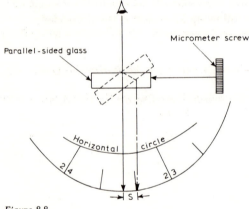

*Figure 8.8*

Assuming the observer's line of sight passes at 90° through the parallel plate glass, the reading is 23° 20′ + S. The parallel plate is rotated using the micrometer screw until an *exact* reading (23° 20′) is obtained on the main scale, as a result of the line of sight being refracted towards the normal and emerging on a parallel path. The distance S, through which the viewer's image was displaced, is recorded on the micrometer scale (11′ 40″) and is a function of the rotation of the plate. Thus it can be seen that rotating the micrometer screw in no way affects the pointing of the theodolite, but back-sets the reading so that rotation of the theodolite is through the total angle of 20° 36′ 20″.

As practically all setting-out work involves the use of the theodolite and/or level, the user should be fully conversant with the various error sources and their effects, as well as the methods of adjustment. Information to this end is available in Chapters 2 and 4.

The use of co-ordinates is now universally applied to the setting out of pipelines, motorways, general road works, power stations, offshore piling and jetty works, housing and high-rise buildings, etc. Thus it can be seen that although the project may vary enormously from site to site the actual setting out is completed using the basic measurements of angle and distance.

There are many advantages to the use of co-ordinates, the main one being that the engineer can set out any part of the works as an individual item, rather than wait for the overall establishment of a setting-out grid.

## 8.4 USE OF GRIDS

Many structures in civil engineering consist of steel or reinforced concrete columns supporting floor slabs. As the disposition of these columns is inevitably that they are at right-angles to each other, the use of a grid, where the grid intersections define the position of the columns, greatly facilitates setting out. It is possible to define several grids as follows.

(1) *Survey grid:* the rectangular co-ordinate system on which the original topographic survey is carried out and plotted (*Figure 8.9*).

(2) *Site grid:* defines the position and direction of the main building lines of the project, as shown in *Figure 8.9*. The best position for such a grid can be determined by simply moving a tracing of the site grid over the original plan so that its best position can be located in relation to the orientation of the major units designed thereon.

In order to set out the site grid, it may be convenient to translate the co-ordinates of

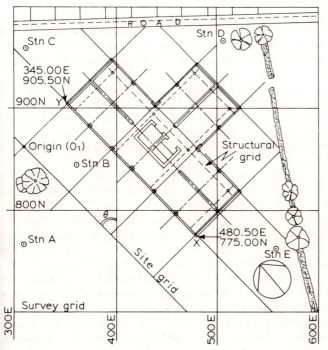

*Figure 8.9*

the site grid to those of the survey grid using the well-known transformation formula

$$E = \Delta E + E_1 \cos \theta - N_1 \sin \theta$$

$$N = \Delta N + N_1 \cos \theta + E_1 \sin \theta$$

where    $\Delta E, \Delta N$ = difference in easting and northing of the respective grid origins
$E_1, N_1$ = the co-ordinates of the point on the site grid
$\theta$ = relative rotation of the two grids
$E, N$ = the co-ordinates of the point transformed to the survey grid

Thus, selected points, say $X$ and $Y$ (*Figure 8.9*) may have their site-grid co-ordinate values transformed to that of the survey grid and so set-out by polars or intersection from the survey control. Now, using $XY$ as a base-line, the site grid may be set out using theodolite and steel tape, all angles being turned off on both faces and grid intervals carefully fixed using the steel tape under standard tension.

When the site grid has been established, each line of the grid should be carefully referenced to marks fixed clear of the area of work. As an added precaution, these marks could be further referenced to existing control or permanent, stable, on-site detail.

(3) *Structural grid*: used to locate the position of the structural elements within the structure and is physically established usually on the concrete floor slab (*Figure 8.9*).

## 8.5 SETTING OUT BUILDINGS

For buildings with normal strip foundations the corners of the external walls are established by pegs located direct from the survey control or by measurement from the site grid. As these pegs would be disturbed in the initial excavations their positions are transferred by theodolite on to profile boards set well vlear of the area of disturbance (*Figure 8.10*). Prior to this their positions must be checked by measuring the diagonals as shown in *Figure 8.11*.

The profile boards must be set horizontal with their top edge at some pre-determined level such as *damp proof course* (DPC) or *finished floor level* (FFL). Wall widths, foundation widths, etc. can be set out along the board with the aid of a steel tape and their positions defined by saw-cuts. They are arranged around the building as shown in *Figure 8.11*. Strings stretched between the appropriate marks clearly define the line of construction.

In the case of buildings constructed with steel or concrete columns, a structural grid must be established to an accuracy of about $\pm 2$ to 3 mm, otherwise the prefabrication beams and steelwork will not fit together without some distortion.

The position of the concrete floor slab may be established in a manner already described. Thereafter the structural grid is physically established by hilty nails or small steel plates set into the concrete. Due to the accuracy required a 1″ theodolite and standardized steel tape corrected for temperature and tension should be used.

Once the bases for the steel columns have been established, the axes defining the centre of each column should be marked on and, using a template oriented to these axes, the positions of the hooding-down bolts defined (*Figure 8.12*). A height mark should be established, using a level, at a set distance (say, 75 mm) below the underside of the base-plate, and this should be constant throughout the structure. It is important that the base-plate starts from a horizontal base to ensure verticality of the column.

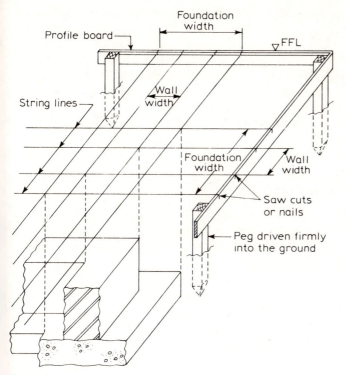

Profile board

Foundation width

FFL

Wall width

String lines

Foundation width

Wall width

Saw cuts or nails

Peg driven firmly into the ground

*Figure 8.10*

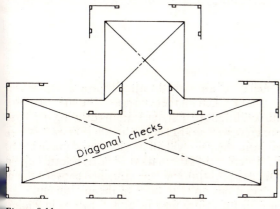

Diagonal checks

*Figure 8.11*

## 8.6 CONTROLLING VERTICALITY

### 8.6.1 Using a plumb bob

In low-rise construction a heavy plumb bob (5 to 10 kg) may be used as shown in *Figure 8.13*. If the external wall was perfectly vertical then, when the plumb bob coincides with the centre of the peg, distance *d* at the top level would equal the offset distance of the peg

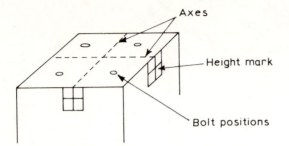

*Figure 8.12*

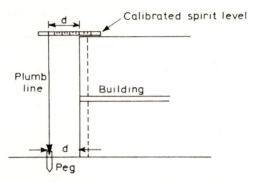

*Figure 8.13*

at the base. This concept can be used internally as well as externally, provided that holes and openings are available.

## 8.6.2 Using a theodolite

If two centre-lines at right-angles to each other are carried vertically up a structure as it is being built, accurate measurement can be taken off these lines and the structure as a whole will remain vertical. Where site conditions permit, the stations defining the 'base figure' (four per line) are placed in concrete well clear of construction (*Figure 8.14(a)*). Lines stretched between marks fixed from the pegs will allow offset measurements to locate the base of the structure. As the structure rises the marks can be transferred up on to the walls by theodolite, as shown in *Figure 8.14(b)*, and lines stretched between them. It is important that the transfer is carried out on both faces of the instrument.

Where the structure is circular in plan the centre may be established as in *Figure 8.14(a)* and the radius swung out from a pipe fixed vertically at the centre. As the structure rises the central pipe is extended by adding more lengths. Its verticality is checked by two theodolites (as in *Figure 8.14(b)*) and its rigidity ensured by supports fixed to scaffolding.

The vertical pipe may be replaced by laser beam or autoplumb, but the laser would still need to be checked for verticality by theodolites.

Steel and concrete columns may also be checked for verticality using the theodolite. By string lining through the columns, positions A–A and B–B may be established for the theodolite (*Figure 8.15*); alternatively, appropriate offsets from the structural grid

(a)

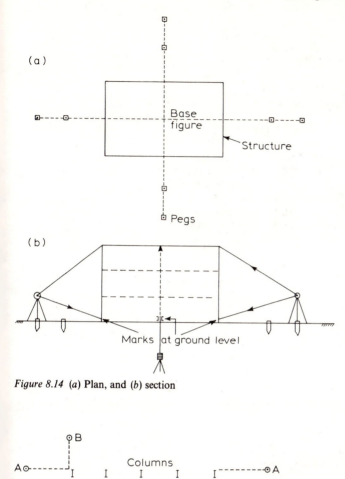

Figure 8.14 (a) Plan, and (b) section

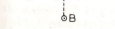

Figure 8.15

lines may be used. With instrument set up at *A*, the outside face of all the uprights should be visible. Now cut the outside edge of the upright at ground level with the vertical hair of the theodolite. Repeat at the top of the column. Now depress the telescope back to ground level and make a fine mark, the difference between the mark and the outside edge of the column is the amount by which the column is out of plumb. Repeat on the opposite face of the theodolite. The whole procedure is now carried out at *B*. If the difference exceeds the specified tolerances the column will need to be corrected.

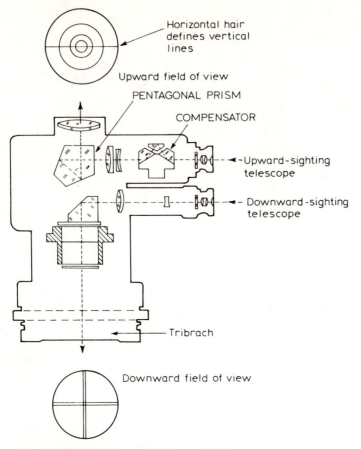

*Figure 8.16* The optical system of the autoplumb

## 8.6.3 Using optical plumbing

For high-rise building the instrument most commonly used is an autoplumb (*Figure 8.16*). This instrument provides a vertical line of sight to an accuracy of ±1 second of arc (1 mm in 200 m). Any deviation from the vertical can be quantified and corrected by rotating the instrument through 90° and observing in all four quadrants; the four marks obtained would give a square, the diagonals of which would intersect at the correct centre point.

A base figure is established at ground level from which fixing measurements may be taken. If this figure is carried vertically up the structure as work proceeds, then identical fixing measurements from the figure at all levels will ensure verticality of the structure (*Figure 8.17*).

To fix any point of the base figure on an upper floor, a Perspex target is set over the opening and the centre point fixed as above. Sometimes these targets have a grid etched on them to facilitate positioning of the marks.

The base figure can be projected as high as the eighth floor, at which stage the finishing trades enter and the openings are closed. In this case the uppermost figure is

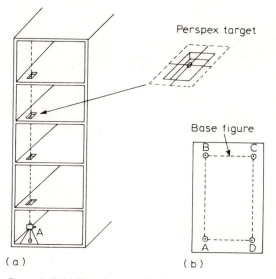

Perspex target

Base figure

( a )                    ( b )

*Figure 8.17* (a) Elevation, and (b) plan

carefully referenced, the openings filled, then the base figure re-established and projected upwards as before.

The shape of the base figure will depend upon the plan shape of the building. In the case of a long rectangular structure a simple base line may suffice but *T* shapes and *Y* shapes are also used.

## 8.7 CONTROLLING GRADING EXCAVATION

This type of setting out generally occurs in drainage schemes where the trench, bedding material and pipes have to be laid to a specified design gradient. Manholes (MH) will need to be set out at every change of direction or at least every 100 m on straight runs. The MH (or inspection chambers) are generally set out first and the drainage courses set out to connevt into them.

The centre peg of the MH is established in the usual way and referenced to four pegs, as in *Figure 8.2*. Alternatively, profile boards may be set around the MH and its dimensions marked on them. If the boards are set out at a known height above formation level the depth of excavation can be controlled, as in *Figure 8.18*.

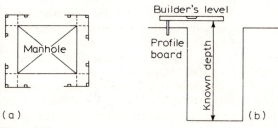

Builder's level

Profile board

Known depth

Manhole

( a )                    ( b )

*Figure 8.18*

## 8.7.1 Use of sight rails

Sight rails (SR) are basically horizontal rails set a specific distance apart and to a specific level such that a line of sight between them is at the required gradient. Thus they are used to control trench excavation and pipe gradient without the need for constant professional supervision.

*Figure 8.19* illustrates SR being used in conjunction with a boning rod (or traveller) to control trench excavation to a design gradient of 1 in 200 (rising). Pegs *A* and *B* are offset a known distance from the centre-line of the trench and levelled from a nearby TBM.

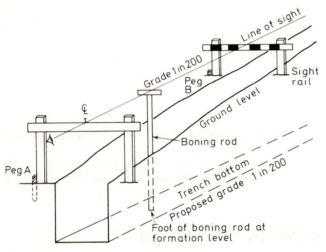

Figure 8.19

Assume peg *A* has a level of 40 m and the formation level of the trench at this point is to be 38 m. It is decided that a reasonable height for the SR above ground would be 1.5 m, i.e. at a level of 41.5; thus the boning rod must be made $(41.5 - 38) = 3.5$ m long, as its cross-head must be on level with the SR when its toe is at formation level.

Consider now peg *B*, with a level of 40.8 m at a horizontal distance of 50 m from *A*. The proposed gradient is 1 in 200, which is 0.25 m in 50 m, thus the formation level at *B* is 38.25 m. If the boning rod is 3.5 m, the SR level at *B* is $(38.25 + 3.5) = 41.75$ m and is set $(41.75 - 40.8) = 0.95$ m above peg *B*. The remaining SRs are established in this way and a line of sight or string stretched between them will establish the trench gradient 3.5 m above the required level. Thus, holding the boning rod vertically in the trench will indicate, relative to the sight rails, whether the trench is too high or too low.

Where machine excavation is used the SR are as in *Figure 8.20*, and offset to the side of the trench opposite to where the excavated soil is deposited.

Knowing the bedding thickness, the invert pipe level may be calculated and a second cross-head added to the boning rod to control the pipe laying, as shown in *Figure 8.21*.

Due to excessive ground slopes it may be necessary to use double sight rails with various lengths of boning rod as shown in *Figure 8.22*.

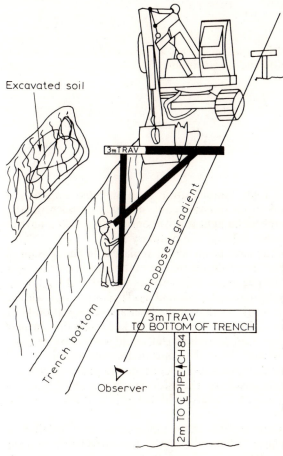

Excavated soil

3mTRAV

Proposed gradient

Trench bottom

3mTRAV
TO BOTTOM OF TRENCH

2m TO ℄ PIPE CH 84

Observer

*Note:*
1 SR offset far enough to allow the machine to pass
2 Offset distance marked on pegs which support SR boards
3 Length of traveller marked on both SR and traveller
4 All crossheads should be levelled with a spirit level
5 Colouring SR and traveller to match can overcome problem of using wrong traveller
6 SR must be offset square to the points to which they refer

*Figure 8.20* Use of offset sight rails (SR)

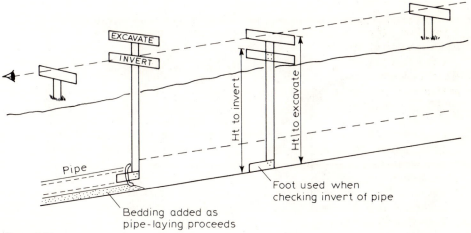

EXCAVATE

INVERT

Ht to invert

Ht to excavate

Pipe

Foot used when checking invert of pipe

Bedding added as pipe-laying proceeds

*Figure 8.21*

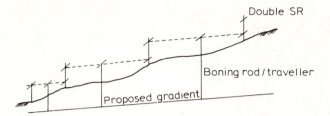

Figure 8.22

## 8.7.2 Use of lasers

The word *laser* is an acronym for Light Amplification by Stimulated Emission of Radiation and is the name applied to an intense beam of highly monochromatic, coherent light. Because of its coherence the light can be concentrated into a narrow beam and will not scatter and become diffused like ordinary light.

In controlling trench excavation the laser beam simply replaces the line of sight or string in the SR situation. It can be set up on the centre-line of the trench, over a peg of known level, and its height above the peg measured to obtain the reduced level of the beam. The instrument is then set to the required grade and used in conjunction with an extendable traveller set to the same height as that of the laser above formation level. When the trench is at the correct level, the laser spot will be picked up on the centre of the traveller target, as shown in *Figure 8.23*. A levelling staff could just as easily replace the traveller, the laser spot being picked up on the appropriate staff reading.

Where machine excavation is used the beam can be picked up on a photo-electric cell fixed at the appropriate height on the machine. The information can be relayed to a console within the cabin, which informs the operator whether he is too high or too low (*Figure 8.24*).

At the pipe-laying stage, a target may be fixed in the pipe and the laser installed on the centre-line in the trench. The laser is then oriented in the correct direction (by bringing it on to a centre-line peg, as in *Figure 8.25*) and depressed to the correct grade

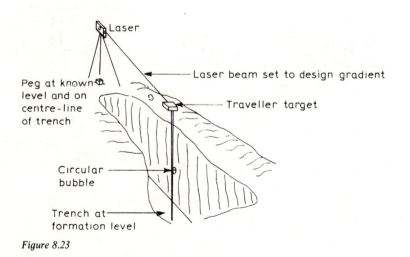

Figure 8.23

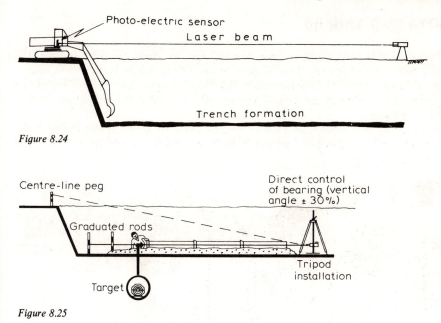

*Figure 8.24*

*Figure 8.25*

of the pipe. A graduated rod, or appropriately-marked ranging pole, can also be used to control formation and sub-grade level (*Figure 8.25*). For large-diameter pipes the laser is mounted inside the pipe using horizontal compression bars.

Where the MH is constructed, the laser can be oriented from within using the system illustrated in *Figure 8.26*. The centre-line direction is transferred down to peg *B* from peg *A* and used to orient the direction of the laser beam.

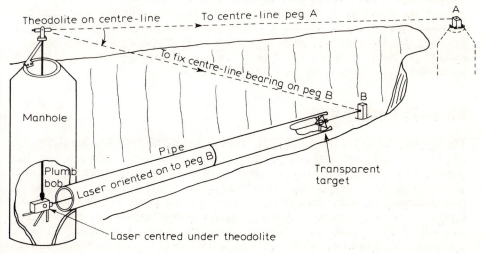

*Figure 8.26*

## 8.8 ROTATING LASERS

Rotating lasers are instruments which are capable of being rotated in both the horizontal and vertical planes, thereby generating reference planes or datum lines.

When the laser is established in the centre of a site over a peg of known level, and at a known height above the peg, a datum of known reduced level is permanently available throughout the site.

Using a vertical staff fitted with a photo-electric detector, levels at any point on the site may be instantly obtained and set out, both above and below ground, as illustrated in *Figure 8.27.*

*Figure 8.27* 1—Rotating laser; 2—Laying sub-grade to laser control; 3—Checking formation level; 4—Fixing wall levels; 5—Taking ground levels; 6—Staff with photoelectric detector fixing foundation levels; 7—Laser plane of reference

Since the laser reference plane covers the whole working area, photo-electric sensors fitted at an appropriate height on earthmoving machinery enables whole areas to be excavated and graded to requirements by the machine operator alone.

Other uses of the rotating laser are illustrated in *Figure 8.28.*

From the above applications it can be seen that basically the laser supplies a reference line at a given height and gradient, and a reference plane similarly disposed. Realizing this, the user may be able to utilize these properties in a wide variety of setting-out situations such as off-shore channel dredging, tunnel guidance (as mentioned in the previous chapter), shaft sinking, etc.

## 8.9 LASER HAZARDS

The potential hazard in the use of lasers is eye damage. There is nothing unique about this form of radiation damage; it can also occur from other, non-coherent, light emitted, for example, by the Sun, arc lamps, projector lamps and other high-intensity sources. If one uses a magnifying lens to focus the Sun's rays on to a piece of paper, the heat generated by such concentration will cause the paper to burst into flames. Similarly with a laser producing a concentrated, powerful beam of light which the eye's lens will further concentrate by focusing it on the retina, thus causing an almost microscopic burn or blister which can cause temporary or permanent blindness. When the beam is focused on the macula (critical area of the retina) serious damage can result.

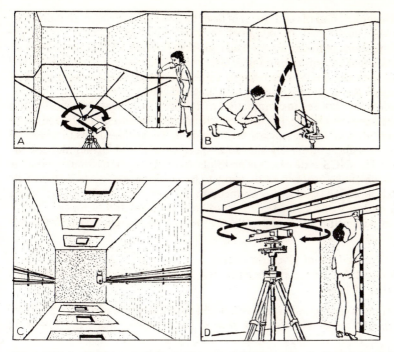

*Figure 8.28* A—Height control; B—Setting out of dividing walls; C—Use of the vertical beam for control of elevator guide rails, and slip-forming structures; D—Setting out and control of suspended ceiling

Since there is no pain or discomfort with a laser burn, the injury may occur several times before vision is impaired. A further complication in engineering surveying is that the beam may be acutely focused through the lens system of a theodolite or other instrument, or may be viewed off a reflecting or refracting surface. It is thus imperative that a safety code be adopted by all personnel involved with the use of lasers.

Under the Health and Safety at Work Act (1974) most sites will be required to adopt the recommendations of the British Standards Institution guide to the safe use of lasers (BS4803). The BSI classifies five types of laser, but only three of these are relevant to on-site working:

Class 2   A visible radiant power of 1 mW. Eye protection is afforded by the blink-reflex mechanism.

Class 3A  Has a maximum radiant power of 1 to 5 mW, with eye protection afforded by the blink-reflex action.

Class 3B  Has a maximum power of 1 to 500 mW. Eye protection is not afforded by blink-reflex. Direct viewing, or viewing of specular reflections, is highly dangerous.

For surveying and setting-out purposes the BSI recommends the use of Classes 2 and 3A only. Class 3B may be used outdoors, if the more stringent safety precautions recommended are observed.

The most significant recommendation of the BSI document is that on sites where lasers are in use there should be a laser safety officer (LSO), who the document defines as 'one who is knowledgeable in the evaluation and control of laser hazards, and has the responsibility for supervision of the control of laser hazards'. Whilst such an individual is not specifically mentioned in conjunction with the Class 2 laser, the legal implications of eye damage might render it advisable to have an LSO present. Such an individual would not only require training in laser safety and law, but would need to be fully conversant with the RICS' Laser Safety Code produced by a working party of certain members of the Royal Institution of Chartered Surveyors.

The RICS code was produced in conjunction with BS4803 and deals specifically with the helium–neon gas laser (He–Ne) as used on site. Whilst the manufacturers of lasers will no doubt comply with the classifications laid down, the modifications to a laser by mirrors or telescopes may completely alter such specifications and further increase the hazard potential. The RICS code presents methods and computations for assessing the possible hazards which the user can easily apply to his working laser system, in both its unmodified and modified states. Recommendations are also made about safety procedures relevant to a particular system from both the legal and technical aspects. The information within the RICS code enables the user to compute such important parameters as

(1) The safe viewing time at given distances.
(2) The minimum safe distance at which the laser source may be viewed directly, for a given period of time.

Such information is vital to the organization and administration of a 'laser site' from both the health and legal aspects, and should be combined with the following precautions:

(1) Ensure that all personnel, random visitors to the site, and where necessary members of the public, are aware of the presence of lasers and the potential eye damage involved.
(2) Using the above-mentioned computations, erect safety barriers around the laser with a radius greater than the minimum safe viewing distance.
(3) Issue laser safety goggles where appropriate.
(4) Avoid, wherever possible, the need to view the laser through theodolites, levels or binoculars.
(5) Where possible, position the laser either well above or well below head height.

## 8.10 ROUTE LOCATION

*Figure 8.29* shows a stretch of route location for a road or railway. In order to control the construction involved, the pegs and profile boards shown must be set out at intervals of 10 to 30 m along the whole stretch of construction.

The first pegs located would be those defining the centre-line of the route (peg E), and the methods of locating these on curves has been dealt with in Chapter 6. The straights would be aligned between adjacent tangent points.

The shoulder pegs C and D, defining the road/railway width, can be set out by appropriate offsets at right-angles to the centre-line chords.

Pegs A and B which define the toe of the embankment (fill) or top edge of the cutting are called *slope stakes*. The side widths from the centre-line are frequently calculated and shown on the design drawings or computer print-outs of setting-out data. This information should be used only as a rough check or guide; the actual location of the slope stake pegs should always be carried out in the field, due to the probable change in

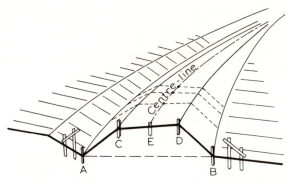

*Figure 8.29*

ground levels since the information was initially compiled. These pegs are established along with the centre-line pegs and are necessary to define the area of top-soil strip.

### 8.10.1 Setting out slope stakes

In *Figures 8.30(a)* and *(b)*, points A and B denote the positions of pegs known as *slope stakes* which define the points of intersection of the actual ground and the proposed side slopes of an embankment or cutting. The method of establishing the positions of the stakes is as follows:

(1) Set up the level in a convenient position which will facilitate the setting out of the maximum number of points therefrom.
(2) Obtain the height of the plane of collimation (HPC) of the instrument by backsighting on to the nearest TBM.
(3) Foresight on to the staff held where it is thought point A may be and obtain the ground level there.
(4) Subtract 'ground level' from 'formation level' and multiply the difference by $N$ to give horizontal distance $x$.
(5) Now tape the horizontal distance $(x + b)$ from the centre-line to the staff. If the *measured distance* to the staff equals the *calculated distance* $(x + b)$, then the staff position is the slope stake position. If not, the operation is repeated with the staff in a different position *until the measured distance agrees with the calculated distance*.

The above 'trial-and-error' approach should always be used on site to avoid errors of scaling the positions from a plan, or accepting, without checking, a computer print-out of the dimensions.

For example, if the side slopes of the proposed embankment are to be 1 vertical, 2 horizontal, the formation level 100 m OD and the ground level at A, say, 90.5 m OD. Then $x = 2(100 - 90.5) = 19$ m, and if the formation width = 20 m, then $b = 10$ m and $(x + b) = 29$ m. Had the staff been held at $A_1$ (which had exactly the same ground level as A) then obviously the calculated distance $(x + b)$ would not agree with the measured distance from centre-line to $A_1$. They would agree only when the staff arrived at the slope stake position A, as $x$ is dependent upon the level at the toe of the embankment, or top of the cutting.

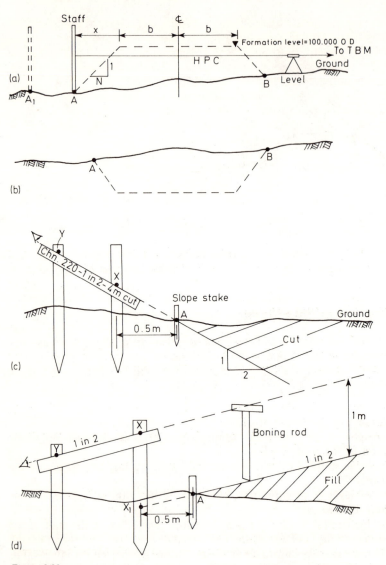

Figure 8.30

## 8.10.2 Controlling earthworks

Batter boards, or *slope rails* as they are sometimes called, are used to control the construction of the side slopes of a cutting or embankment (see *Figures 8.30(c)* and (*d*), and *8.29*).

Consider *Figure 8.30(c)*, if the stake adjacent to the slope stake is set 0.5 m away, then, for a grade of 1 vertical to 2 horizontal, the level of point *X* will be 0.25 m higher than the ground level at A. From *X*, a batter board is fixed at a grade of 1 in 2, using a 1 in 2 template and a spirit level. Stakes *X* and *Y* are usually no more than 1 m apart. Information such as chainage, slope and depth of cut are marked on the batter board.

    In the case of an embankment, *Figure 8.30(d)*, a boning rod is used in the control of the slope. Assuming that a boning rod 1 m high is to be used, then as the near stake is, say, 0.5 m from the slope stake, point $x_1$ will be 0.25 m lower than the ground level at A, hence point $x$ will be 0.75 m above the ground level of A. The batter board is then fixed from $x$ in a similar manner to that already described.

The formation and sub-base, which usually have setting-out tolerances in the region of $\pm 25$ mm, can be located with sufficient accuracy using profiles and travellers. *Figure 8.31* shows the use of triple profiles for controlling camber, whilst different lengths of traveller will control the thickness required.

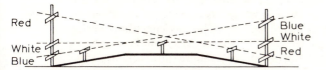

Figure 8.31

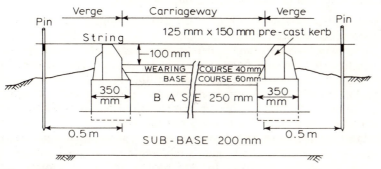

Figure 8.32

    Laying of the base course (60 mm) and wearing course (40 mm) calls for much smaller tolerances, and profiles are not sufficiently accurate; the following approach may be used:

    Pins or pegs are established at right-angles to the centre-line at about 0.5 m beyond the kerb face (*Figure 8.32*). The pins or pegs are accurately levelled from the nearest TBM and a coloured tape placed around them at 100 mm above finished road level; this will be at the same level as the top of the kerb. A cord stretched between the pins will give kerb level, and with a tape the distances to the top of the sub-base, top of the base course and top of the wearing course, can be accurately fixed (or dipped).

    The distance to the kerb face can also be carefully measured in from the pin in order to establish the kerb line. This line is sometimes defined with further pins and the level of the kerb top marked on.

## 8.11 RESPONSIBILITY ON SITE

Responsibility with regard to setting out is defined in Clause 17 of the ICE Conditions of Contract:

> The contractor shall be responsible for the true and proper setting out of the works, and for the correctness of the position, levels, dimensions, and alignment of all parts of the works, and for the provision of all necessary instruments, appliances, and labour in connection therewith. If, at any time during the progress of the works, any error shall appear or arise in the position, levels, dimensions, or alignment of any part of the works, the contractor, on being required so to do by the engineer, shall, *at his own cost*, rectify such error to the satisfaction of the engineer, unless such error is based on incorrect data supplied in writing by the engineer or the engineer's representative, in which case the cost of rectifying the same shall be borne *by the employer*. The checking of any setting out, or of any line or level, by the engineer or the engineer's representative, shall not, in any way, relieve the contractor of *his* responsibility for the correctness thereof, and the contractor shall carefully protect *and preserve* all bench-marks, sight rails, pegs, and other things used in setting out the works.

The clause specifies three persons involved in the process, namely, the employer, the engineer and the agent, whose roles are as follows:

The *employer*, who may be a government department, local authority or private individual, requires to carry out and finance a particular project. To this end, he commissions an *engineer* to investigate and design the project, and to take responsibility for the initial site investigation, surveys, plans, designs, working drawings, and setting-out data. On satisfactory completion of his work he lets the contract to a contractor whose duty it is to carry out the work.

On site the employer is represented by the engineer or his representative, referred to as the *resident engineer* (RE), and the contractor's representative is called the *agent*.

The engineer has overall responsibility for the project and must protect the employer's interest without bias to the contractor. The agent is responsible for the actual construction of the project.

## 8.12 RESPONSIBILITY OF THE SETTING-OUT ENGINEER

The setting-out engineer should establish such a system of work on site that will ensure the accurate setting out of the works well in advance of the commencement of construction. To achieve this, the following factors should be considered:

(1) A complete and thorough understanding of the plans, working drawings, setting-out data, tolerances involved and the time scale of operations. Checks on the setting-out data supplied should be immediately implemented.

(2) A complete and thorough knowledge of the site, plant and relevant personnel. Communications between all individuals is vitally important. Field checks on the survey control already established on site, possibly by contract surveyors, should be carried out at the first opportunity.

(3) A complete and thorough knowledge of the survey instrumentation available on site, including the effect of instrumental errors on setting-out observations. At the first opportunity, a base should be established for the calibration of tapes, EDM equipment, levels and theodolites.

(4) A complete and thorough knowledge of the stores available, to ensure an adequate and continuing supply of pegs, pins, chalk, string, paint, timber, etc.
(5) Office procedure should be so organized as to ensure easy access to all necessary information. Plans should be stored flat in plan drawers, and those amended or superseded should be withdrawn from use and stored elsewhere. Field and level books should be carefully referenced and properly filed. All setting-out computations and procedures used should be clearly presented, referenced and filed.
(6) Wherever possible, independent checks of the computation, abstraction, and extrapolation of setting-out data and of the actual setting-out procedures should be made.

It can be seen from this brief itinerary of the requirements of a setting-out engineer, that such work should never be allocated, without complete supervision, to junior, inexperienced members of the site team.

# Appendix A (Chapter 3)

# Earthworks program

```
10 REM: CUT AND FILL CALCULATIONS
20 REM: BY BRIAN MERRONY, SCHOOL OF CIVIL ENGINEERING
30 REM: KINGSTON POLYTECHNIC
40 DIM GY(50) GX(50),RY(50),RX(50),S(2)
50 L$="---- ----------------------------" \ PRINT L$
60 PRINT"EARTHWORKS" \ PRINT L$
70 REM----- START OF SECTION DATA
80 INPUT"FILL SLOPE ";FS
90 INPUT"CUT SLOPE " CS
100 INPUT "NO OF GROUND POINTS ";NG
110 PRINT"ENTER GROUND OFFSETS AND LEVELS (IN PAIRS)"
120 FOR I=0 TO NG-1
130 PRINT"POINT";I+1; \ INPUT GX(I),GY(I)
140 NEXT I \ PRINT L$
150 INPUT "NO OF ROAD POINTS ";NR
160 PRINT"ENTER ROAD OFFSETS AND LEVELS (IN PAIRS)"
170 FOR I=1 TO NR
180 PRINT"POINT";I; \ INPUT RX(I),RY(I)
190 NEXT I \ PRINT L$
200 REM ***** CHECK THAT GROUND SPANS ROAD *****
210 IF GX(0)<=RX(1) AND GX(NG-1)>=RX(NR) THEN 230
220 PRINT "GROUND DOES NOT SPAN ROAD" \ GOTO 1110
230 REM *****TREAT LEFT HAND SIDE *****
240 P=0
250 P=P+1 \ IF GX(P)<RX(1) THEN 250
260 A=1 \ B=0 \ Q=1 \ GOSUB 940
270 REM *****TREAT RIGHT HAND SIDE*****
280 P=NG-1
290 P=P-1 \ IF GX(P)>RX(NR) THEN 290
300 A=-1 \ B=NG-1 \ Q=NR \ GOSUB 940
310 REM ***** START OF MAIN PASS *****
320 P=0 \ Q=0 \ F=1
330 AA=GY(0) \ BB=RY(0) \ HH=ABS(AA-BB)
340 A=1 \ IF AA<BB THEN A=2
350 EE=GX(0)
360 REM----- START OF LOOP
370 P=P+1 \ Q=Q+1 \ X1=EE
380 RX=RX(Q) \ GX=GX(P)
390 IF GX>RX THEN 470
400 IF GX=RX THEN 530
410 REM *****CONDITION GX<RX*****
420 FC=GX \ CC=GY(P)
430 Y1=BB \ Y2=RY(Q) \ X2=RX \ XX=GX
440 DD=(Y2-Y1)*(XX-X1)/(X2-X1)+Y1
450 Q=Q-1
460 GOTO 560
470 REM *****GX>RX CONDITION*****
480 FG=RX \ DD=RY(Q)
490 Y1=AA \ Y2=GY(P) \ X2=GX \ XX=RX
500 CC=(Y2-Y1)*(XX-X1)/(X2-X1)+Y1
510 P=P-1
520 GOTO 560
530 REM *****GX=RX CONDITION*****
540 FG=GX \ CC=GY(P) \ DD=RY(Q)
550 Y1=AA \ Y2=CC \ X2=GX
560 REM----- EVALUATE AREAS
570 B=1 \ IF CC<DD THEN B=2
580 II=ABS(CC-DD)
590 GG=FG-EE
600 IF A=B THEN 750
610 REM----- TWO TRIANGLES
620 Y=GG*HH/(HH+II)
630 XX=EE+Y
640 JJ=Y*HH/2
650 IF F=1 THEN F=0 \ JJ=0
660 S(A)=S(A)+JJ
670 S3=S3+S(1) \ S(1)=0
680 S4=S4+S(2) \ S(2)=0
690 KK=GG*II*II/(2*(HH+II))
700 S(B)=S(B)+KK
710 Y=(Y2-Y1)*(XX-X1)/(X2-X1)+Y1
720 PRINT "INTERSECTION"
730 PRINT XX Y
740 GOTO 770
750 REM ---- TRAPEZIUM
760 IF F=0 THEN S(A)=S(A)+(HH+II)*GG/2
770 REM ----TEST FOR END OF CROSS-SECTION
780 IF P=NG-1 AND Q=NR+1 THEN 810
790 AA=CC \ BB=DD \ EE=FG \ HH=II \ A=B
800 GOTO 360
810 REM -----CALCULATE VOLUMES AND UPDATE FOR NEXT SECTION
820 PRINT"CUT AREA, FILL AREA" \ PRINT S3 S4
830 S(1)=0 \ S(2)=0
840 V1=(S5+S3)*.5*LL \ V2=(S6+S4)*.5*LL
850 S5=S3 \ S6=S4 \ S3=0 \ S4=0
860 PRINT"CUT VOLS, INCREMENT AND TOTAL"
870 V3=V3+V1 \ PRINT V1,V3
880 PRINT"FILL VOLS, INCREMENT AND TOTAL"
890 V4=V4+V2 \ PRINT V2,V4 \ PRINT L$
900 PRINT"CHAINAGE INTERVAL (INSERT ZERO TO TERMINATE RUN)"
910 INPUT LL
920 IF LL=0 THEN 1110
930 GOTO 70
940 REM -----DECIDE CUT OR FILL AND EXTRAPOLATE IF NECESSARY
950 XX=RX(Q)
960 Y1=GY(P-A) \ X1=GX(P-A) \ Y2=GY(P) \ X2=GX(P)
970 Y=(Y2-Y1)*(XX-X1)/(X2-X1)+Y1
980 F=0 \ IF RY(Q)<Y THEN F=1
990 IF RY(Q)=Y THEN RY(Q)=Y+0.0001
1000 S=-FS \ IF F=1 THEN S=CS
1010 Y1=GY(B) \ X1=GX(B) \ Y2=GY(B+A) \ X2=GX(B+A)
1020 XX=GX(B)
1030 Y=RY(Q)+ABS(RX(Q)-XX)*S
1040 IF (F=0 AND GY(B)>Y) OR (F=1 AND GY(B)<Y) THEN 1100
1050 XX=XX-100*A \ Y=Y+100*S
1060 GY(B)=(Y2-Y1)*(XX-X1)/(X2-X1)+Y1
1070 GX(B)=XX
1080 IF (F=0 AND GY(B)>Y) OR (F=1 AND GY(B)<Y) THEN 1100
1090 PRINT "EXTRAPOLATION FAILED" \ GOTO 1110
1100 RX(Q-A)=GX(B) \ RY(Q-A)=Y \ RETURN
1110 END
```

# EARTHWORKS PROGRAM: BASIC

The program is used for earthwork computation and mass-haul data.

From offsets and levels of existing ground cross-sections and proposed road cross-sections at successive chainage points along the centre-line of a route location; cut and fill volumes are computed, accumulated volumes for mass haul produced and the chainage and level of the intersection of the side slopes and existing ground are found.

## Input data (the program is interactive and will request input data)

(1) Side slope in FILL; e.g. for 1 in 2, enter 0.5.

(2) Side slope in CUT; this may be the same as or different from that adopted for fill.

(3) The number of points defining the EXISTING GROUND in the cross-section.

(4) The distance from the centre-line and level of the GROUND POINTS. Distances to the *left* of the centre-line are *negative*, those to the *right* are *positive*. *N.B.* They are entered into the computer IN SEQUENCE FROM THE EXTREME LEFT TO RIGHT, of the cross-section.

(5) The number of points defining the proposed ROAD formation.

(6) As in (4), but for road points.

(7) The chainage interval to the next cross-section. If there are no further cross-sections the entry of a zero chainage interval terminates the run. (Refer example overleaf.)

## Output data

### (1) Ground/road intersection points

The distance from the centre-line and level at which the road side-slopes intersect the existing ground; i.e. the top of the cutting or toe of the embankment (the *slope stake* position).

### (2) Areas

Areas of the cross-sections.

### (3) Volumes

The volume at each cross-section and the accumulated volume. The latter is used for mass haul construction.

An example of the input and output for a hill-side section in cut-and-fill is given overleaf.

```
------------------------------------
EARTHWORKS
------------------------------------
FILL SLOPE ? 0.5                                            ⌐
CUT SLOPE ? 1.0
NO OF GROUND POINTS ? 7
ENTER GROUND OFFSETS AND LEVELS (IN PAIRS)          INPUT DATA OF
POINT 1 ? -30, 100                                  CROSS-SECTION
POINT 2 ? -20, 95.5                                 GROUND LEVELS
POINT 3 ? -10, 102
POINT 4 ? 0, 103
POINT 5 ? 10, 105.6
POINT 6 ? 20, 106.8
POINT 7 ? 30, 106.9
------------------------------------
NO OF ROAD POINTS ? 3
ENTER ROAD OFFSETS AND LEVELS (IN PAIRS)            INPUT DATA OF
POINT 1 ? -18, 103.7                                PROPOSED ROAD
POINT 2 ? 0, 104                                    LEVELS
POINT 3 ? 18, 103.7
------------------------------------
INTERSECTION
-27.5789        98.9105
INTERSECTION
 3.61445        103.94
INTERSECTION
 21.1111        106.811
CUT AREA, FILL AREA
 28.7184        92.2913
CUT VOLS, INCREMENT AND TOTAL
 0              0                                    OUTPUT
FILL VOLS, INCREMENT AND TOTAL
 0              0
------------------------------------
CHAINAGE INTERVAL (INSERT ZERO TO TERMINATE RUN)
? 30
FILL SLOPE ? 0.5
CUT SLOPE ? 1.0
NO OF GROUND POINTS ? 7
ENTER GROUND OFFSETS AND LEVELS (IN PAIRS)
POINT 1 ?

        etc., etc. ...
```

# Appendix B (Chapter 6)

# Derivation of clothoid spiral formulae

$AB$ is an infinitely small portion ($\delta l$) of the transition curve $T_1 t_1$. (*Figure B.1*)

$$\delta x/\delta l = \cos \phi = (1 - \phi^2/2! + \phi^4/4! - \phi^6/6! \ldots)$$

The basic equation for a clothoid curve is: $\phi = l^2/2RL$

$$\therefore \ \delta x/\delta l = \left(1 - \frac{(l^2/2RL)^2}{2!} + \frac{(l^2/2RL)^4}{4!} - \frac{(l^2/2RL)^6}{6!} \ldots\right)$$

Integrating: $\quad x = l\left(1 - \frac{l^4}{40(RL)^2} + \frac{l^8}{3456(RL)^4} - \frac{l^{12}}{599040(RL)^6}\right)$

$$= l - \frac{l^5}{40(RL)^2} + \frac{l^9}{3456(RL)^4} - \frac{l^{13}}{599040(RL)^6} + \cdots$$

when $x = X$, $l = L$ and:

$$X = L - \frac{L^3}{5.4.2!R^2} + \frac{L^5}{9.4^2.4!R^4} - \frac{L^7}{13.4^3.6!R^6} + \cdots \tag{6.23}$$

Similarly, $\quad \delta y/\delta l = \sin \phi = (\phi - \phi^3/3! + \phi^5/5! \ldots)$

Substituting for $\phi$ as previously

$$\delta y/\delta l = \frac{l^2}{2RL} - \frac{(l^2/2RL)^3}{6} + \frac{(l^2/2RL)^5}{120} \cdots$$

Integrating: $\quad y = l\left(\frac{l^2}{6RL} - \frac{l^6}{336(RL)^3} + \frac{l^{10}}{42240(RL)^5} \cdots\right)$

$$= \frac{l^3}{6RL} - \frac{l^7}{336(RL)^3} + \frac{l^{11}}{42240(RL)^5} \cdots$$

when $y = Y$, $l = L$, and

$$Y = \frac{L^2}{3.2R} - \frac{L^4}{7.3!2^3R^3} + \frac{L^6}{11.5!2^5R^5} \cdots \tag{6.24}$$

The basic equation for a clothoid is $l = a(\phi)^{1/2}$, where $l = L$, $\phi = \Phi$ and $L = a(\Phi)^{1/2}$, then squaring and dividing gives

$$L^2/l^2 = \Phi/\phi \tag{6.21}$$

From *Figure B.1*

$$\tan \theta = \frac{y}{x} = \frac{l(l^2/6RL - l^6/336(RL)^3 + l^{10}/42240(RL)^5)}{l(1 - l^4/40(RL)^2 + l^8/3456(RL)^4)}$$

However, $l = (2RL\phi)^{1/2}$

$$\therefore \ \tan \theta = \frac{(2RL\phi)^{1/2}(\phi/3 - \phi^3/42 + \phi^5/1320)}{(2RL\phi)^{1/2}(1 - \phi^2/10 + \phi^4/216)}$$

$$= \left(\frac{\phi}{3} - \frac{\phi^3}{42} + \frac{\phi^5}{1320}\right)\left(1 - \frac{\phi^2}{10} + \frac{\phi^4}{216}\right)^{-1}$$

Let $x = -(\phi^2/10 - \phi^4/216)$ and expanding the second bracket binomially

i.e.   $(1 + x)^{-1} = 1 - nx + \frac{n(n - 1)x^2}{2!} \cdots$

$$= 1 + \frac{\phi^2}{10} - \frac{\phi^4}{216} + \frac{2}{2!}\left(\frac{\phi^2}{10} - \frac{\phi^4}{216}\right)^2$$

$$= 1 + \frac{\phi^2}{10} - \frac{\phi^4}{216} + \frac{\phi^4}{100} + \frac{\phi^8}{81.4!4!} - \frac{2\phi^6}{45.2!4!}$$

$$= 1 + \frac{\phi^2}{10} - \frac{\phi^4}{216} + \frac{\phi^4}{100}$$

$$\therefore \ \tan \theta = \left(\frac{\phi}{3} - \frac{\phi^3}{42} + \frac{\phi^5}{11.5!}\right)\left(1 + \frac{\phi^2}{10} - \frac{\phi^4}{216} + \frac{\phi^4}{100}\right)$$

$$= \frac{\phi}{3} + \frac{\phi^3}{105} + \frac{26\phi^5}{155925}$$

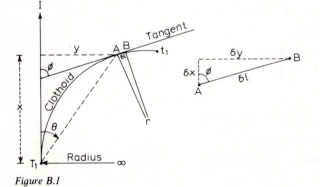

*Figure B.1*

However, as $\quad \theta = \tan \theta - \dfrac{1}{3}\tan^3 \theta + \dfrac{1}{5}\tan^5 \theta$

then $\qquad\qquad = \dfrac{\phi}{3} - \dfrac{8\phi^3}{2835} - \dfrac{32\phi^5}{467775} \cdots$ $\qquad\qquad\qquad\qquad$ (6.20)

$\qquad\qquad\qquad = \dfrac{\phi}{3} - N$

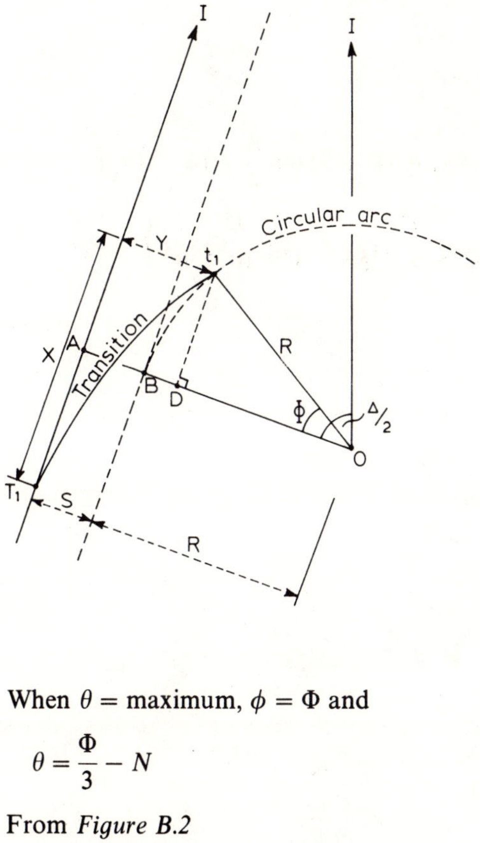

*Figure B.2*

When $\theta = $ maximum, $\phi = \Phi$ and

$$\theta = \dfrac{\Phi}{3} - N$$

From *Figure B.2*

$$BD = BO - DO = R - R\cos \Phi = R(1 - \cos \Phi)$$

$$= R\left(\dfrac{\Phi^2}{2!} + \dfrac{\Phi^4}{4!} + \dfrac{\Phi^6}{6!}\cdots\right)$$

but $\Phi = L/2R$,

$$\therefore \ BD = \dfrac{L^2}{2!2^2 R} + \dfrac{L^4}{4!2^4 R^3} + \dfrac{L^6}{6!2^6 R^5}$$

and $\quad Y = \dfrac{L^2}{3.2R} - \dfrac{L^4}{7.3!2^3 R^3} + \dfrac{L^6}{11.5!2^5 R^5}$

$$\therefore \text{ Shift} = S = (Y - BD) = \frac{L^2}{24R} - \frac{L^4}{3!7.8.2^3R^3} + \frac{L^6}{5!11.12.2^5R^5}$$

From *Figure B.2*

$$Dt_1 = R \sin \phi = R\left(\Phi - \frac{\Phi^3}{3!} + \frac{\Phi^5}{5!} \cdots\right)$$

but $\Phi = L/2R$

$$\therefore \; Dt_1 = \frac{L}{2} - \frac{L^3}{3!2^3R^2} + \frac{L^5}{5!2^5R^4} \cdots$$

also   $X = L - \dfrac{L^3}{5.4.2!R^2} + \dfrac{L^5}{9.4^2.4!R^4} \cdots$

Tangent length $= T_1I = (R + S) \tan \dfrac{\Delta}{2} + AT_1 = (R + S) \tan \dfrac{\Delta}{2} + (X - Dt_1)$

$$= (R + S) \tan \frac{\Delta}{2} + \left(\frac{L}{2} - \frac{L^3}{2!5.6.2^2R^2} + \frac{L^5}{4!9.10.2^4R^4}\right)$$

$$= (R + S) \tan \frac{\Delta}{2} + C \tag{6.18}$$

# Abbreviations

| | |
|---|---|
| AMV | angular momentum vector |
| AOD | above ordnance datum |
| BM | bench mark |
| BS | back sight |
| BSI | British Standards Institution |
| CP | change point |
| DGM | digital ground model |
| DoT | Department of Transport |
| DPC | damp-proof course |
| EDM | electromagnetic distance-measuring |
| FFL | finished floor level |
| FL | face left |
| FOSD | full overtaking sight distance |
| FR | face right |
| FS | fore sight |
| HPC | height of the plane of collimation |
| HWM | high-water mark |
| ICE | Institution of Civil Engineers |
| IS | intermediate sight |
| KP | Kingston Polytechnic |
| LAT | lowest astronomical tide |
| LSO | laser safety officer |
| LU | London University |
| LWM | low-water mark |
| MH | manhole |
| MHD | mass-haul diagram |
| MSL | mean sea level |
| NCB | National Coal Board |
| NG | national grid |
| OD | ordnance datum |
| OS | Ordnance Survey |
| PPM | parallel-plate micrometer |
| PSE | proportional standard error |
| QB | quadrant bearing |
| RE | resident engineer |
| R-and-F | rise-and-fall |
| RL | reduced level |
| RO | reference object |
| RRL | Road Research Laboratory |
| RSM | Royal School of Mines |
| SR | sight rail |
| SSD | stopping sight distance |
| TBM | temporary bench mark |
| TMP | transverse Mercator projection |
| TTS | three-tripod system |
| VC | vertical curve |
| WCB | whole-circle bearing |

# Index